Galois Theory and Its Algebraic Background

SECOND EDITION

Galois theory, the theory of polynomial equations and their solutions, is one of the most fascinating and beautiful subjects in pure mathematics. Using group theory and field theory, it provides a complete answer to the problem of the solubility of polynomial equations by radicals: that is, determining when and how a polynomial equation can be solved by repeatedly extracting roots using elementary algebraic operations.

This textbook contains a fully detailed account of Galois theory and the algebra that it needs, and is suitable for both those following a course of lectures and the independent reader (who is assumed to have no previous knowledge of Galois theory). This second edition has been significantly revised and reordered; the first part develops the basic algebra that is needed, and the second part gives a comprehensive account of Galois theory. There are applications to ruler and compass constructions, and to the solution of classical mathematical problems of ancient times. There are new exercises throughout, and carefully selected examples will help the reader develop a clear understanding of the mathematical theory.

D.J.H. GARLING is Emeritus Reader in Mathematical Analysis at the University of Cambridge and Fellow of St John's College, Cambridge. He has 50 years' experience of teaching undergraduate students and has written several books on mathematics, including *Inequalities: A Journey into Linear Analysis* (Cambridge University Press, 2007) and *A Course in Mathematical Analysis* (three volumes, Cambridge University Press, 2013–2014).

T0179671

Galois Theory and Its Algebraic Background

SECOND EDITION

D.J.H. Garling
University of Cambridge

CAMBRIDGE
UNIVERSITY PRESS

CAMBRIDGE
UNIVERSITY PRESS

University Printing House, Cambridge CB2 8BS, United Kingdom

One Liberty Plaza, 20th Floor, New York, NY 10006, USA

477 Williamstown Road, Port Melbourne, VIC 3207, Australia

314–321, 3rd Floor, Plot 3, Splendor Forum, Jasola District Centre, New Delhi – 110025, India

103 Penang Road, #05–06/07, Visioncrest Commercial, Singapore 238467

Cambridge University Press is part of the University of Cambridge.

It furthers the University's mission by disseminating knowledge in the pursuit of
education, learning, and research at the highest international levels of excellence.

www.cambridge.org
Information on this title: www.cambridge.org/9781108838924
DOI: 10.1017/9781108979184

First edition © Cambridge University Press 1986
Second edition © D.J.H. Garling 2022

First published 1986
Reprinted with corrections 1988, 1991, 1993, 1995
Second edition 2022

Printed in the United Kingdom by TJ Books Limited, Padstow, Cornwall, 2022

A catalogue record for this publication is available from the British Library.

Library of Congress Cataloging-in-Publication Data

Names: Garling, D. J. H., author.
Title: Galois theory and its algebraic background / D.J.H. Garling.
Other titles: Course in Galois theory
Description: Second edition. | Cambridge ; New York, NY : Cambridge
University Press, 2021. | First edition published as Course in Galois
theory, Cambridge University Press, 1986. | Includes bibliographical references and index.
Identifiers: LCCN 2021002526 (print) | LCCN 2021002527 (ebook) |
ISBN 9781108838924 (hardback) | ISBN 9781108969086 (paperback) | ISBN 9781108979184 (epub)
Subjects: LCSH: Galois theory.
Classification: LCC QA214 .G367 2021 (print) — LCC QA214 (ebook) | DDC 512/.32–dc23
LC record available at https://lccn.loc.gov/2021002526
LC ebook record available at https://lccn.loc.gov/2021002527

ISBN 978-1-108-83892-4 Hardback
ISBN 978-1-108-96908-6 Paperback

Contents

Preface   *page* ix

PART I   THE ALGEBRAIC BACKGROUND

1   **Groups**   3
1.1   Groups   3
1.2   Finite Abelian Groups   9
1.3   Finite Permutation Groups   11
1.4   Group Series   17
1.5   Soluble Groups   18
1.6   *p*-Groups and Sylow Theorems   22

2   **Integral Domains**   25
2.1   Commutative Rings with 1   25
2.2   Polynomials   26
2.3   Homomorphisms and Ideals   27
2.4   Integral Domains   29
2.5   Fields and Fractions   31
2.6   The Ordered Set of Ideals in an Integral Domain   35
2.7   Factorization   36
2.8   Unique Factorization   38
2.9   Principal Ideal Domains and Euclidean Domains   41
2.10   Polynomials Over Unique Factorization Domains   44
2.11   More About Fields   47
2.12   Kronecker's Algorithm   48
2.13   Eisenstein's Criterion   50
2.14   Localization   51

3	**Vector Spaces and Determinants**	54
3.1	Vector Spaces	54
3.2	The Infinite-Dimensional Case	60
3.3	Characters and Automorphisms	61
3.4	Determinants	62

PART II   THE THEORY OF FIELDS AND GALOIS THEORY

4	**Field Extensions**	69
4.1	Introduction	69
4.2	Field Extensions	70
4.3	Algebraic and Transcendental Extensions	73
4.4	Algebraic Extensions	77
4.5	Monomorphisms of Algebraic Extensions	80
5	**Ruler and Compass Constructions**	81
5.1	Some Classical Problems	81
5.2	Constructible Points	81
6	**Splitting Fields**	85
6.1	Introduction	85
6.2	Splitting Fields	86
6.3	The Extension of Monomorphisms	89
6.4	Some Examples	94
7	**Normal Extensions**	98
7.1	Basic Properties	98
7.2	Monomorphisms and Automorphisms	101
8	**Separability**	103
8.1	Basic Ideas	103
8.2	Monomorphisms and Automorphisms	104
8.3	Galois Extensions	106
8.4	Differentiation	106
8.5	Inseparable Polynomials	108
9	**The Fundamental Theorem of Galois Theory**	112
9.1	Field Automorphisms, Fixed Fields and Galois Groups	112
9.2	Linear Independence	113
9.3	The Size of a Galois Group is the Degree of the Extension	115
9.4	The Galois Group of a Polynomial	116
9.5	The Fundamental Theorem of Galois Theory	118

10	**The Discriminant**	122
10.1	The Discriminant	122
11	**Cyclotomic Polynomials and Cyclic Extensions**	126
11.1	Cyclotomic Polynomials	126
11.2	Irreducibility	128
11.3	The Galois Group of a Cyclotomic Polynomial	129
11.4	A Necessary Condition	131
11.5	Abel's Theorem	132
11.6	Norms and Traces	134
11.7	A Sufficient Condition	135
11.8	Kummer Extensions	137
12	**Solution by Radicals**	140
12.1	Polynomials with Soluble Galois Groups	140
12.2	Polynomials which are Soluble by Radicals	141
13	**Regular Polygons**	146
13.1	Fermat Primes and Fermat Numbers	146
13.2	Regular Polygons	147
13.3	Constructing a Regular Pentagon	148
14	**Polynomials of Low Degree**	149
14.1	Quadratic Polynomials	149
14.2	Cubic Polynomials	150
14.3	Quartic Polynomials	153
15	**Finite Fields**	156
15.1	Finite Fields	156
15.2	Polynomials in $\mathbb{Z}_p[x]$	157
15.3	Polynomials of Low Degree over a Finite Field	158
16	**Quintic Polynomials**	161
17	**Further Theory**	164
17.1	Simple Extensions	165
17.2	The Theorem of the Primitive Element	166
17.3	The Normal Basis Theorem	168
18	**The Algebraic Closure of a Field**	170
18.1	Introduction	170
18.2	The Existence of an Algebraic Closure	171
18.3	The Uniqueness of an Algebraic Closure	175
18.4	Conclusions	176

19   Transcendental Elements and Algebraic Independence   177
19.1 Transcendental Elements and Algebraic Independence   177
19.2 Transcendence Bases   180
19.3 Transcendence Degree   181
19.4 The Tower Law for Transcendence Degree   182
19.5 Lüroth's Theorem   183

20   Generic and Symmetric Polynomials   186
20.1 Generic and Symmetric Polynomials   186

Appendix: The Axiom of Choice   189
Index   192

Preface

Galois theory is one of the most fascinating and enjoyable branches of algebra. The problems with which it is concerned have a long and distinguished history: the problems of duplicating a cube or trisecting an angle go back to the Greeks, and the problem of solving a cubic, quartic or quintic equation to the Renaissance. Many of the problems that are raised are of a concrete kind (and this, surely, is why it is so enjoyable) and yet the needs of the subject have led to substantial development in many branches of abstract algebra: in particular, in the theory of fields, the theory of groups, the theory of vector spaces and the theory of commutative rings.

In this book, Galois theory is treated as it should be, as a subject in its own right. Nevertheless, in the process, I have tried to show its relationship to various topics in abstract algebra: an understanding of the structures of abstract algebra helps give a shape to Galois theory and conversely Galois theory provides plenty of concrete examples which show the point of abstract theory.

The book comprises two unequal parts. In the first part, an account is given of the algebra that is needed for Galois theory. Much of this may well be familiar to the reader, but is included both for completeness and to introduce the terminology and notation that is used. Much of the algebra (groups, rings, fields and vector spaces) has general interest, and of course the development of Galois theory was responsible for the development of many algebraic ideas. We shall concentrate on presenting those algebraic ideas and results that are needed for Galois theory. For example, it is important to know that in the right circumstances, the factorization of polynomials with coefficients in a ring is essentially unique. Group theory plays a large part in Galois theory, but has developed into a huge subject. We shall concentrate on those parts, such as the theory of soluble groups, which are needed in Galois theory.

The second, more substantial, part is concerned with the theory of fields and with Galois theory, and contains the main material of the book; indeed, many readers may wish to start here and refer back to the first part as necessary. Of its nature, the theory develops an inexorable momentum. Nevertheless, there are many digressions (for example, concerning geometric constructions, finite fields and the solution of cubic and quartic equations): one of the pleasures of Galois theory is that there are many examples which illustrate and depend upon the general theory, but which also have an interest of their own. The high point of the book is of course the resolution of the problem of when a polynomial is solvable by radicals. I have, however, tried to emphasize (in the final chapter in particular) that this is not the end of the story: the resolution of the problem raises many new problems, and Galois theory is still a lively subject.

The last three chapters have a more abstract nature, and require the use of Zorn's lemma. In full generality, in the uncountable case, this depends upon the Axiom of Choice; this is discussed in the Appendix. Algebra is principally concerned with finite operations and relations, and is therefore largely concerned with finite or countable sets, and so these chapters have a rather hybrid quality.

Two hundred exercises are scattered through the text. It has been suggested to me that this is rather few: I think that anyone who honestly tries them all will disagree! In my opinion, textbook exercises are often too straightforward, but some of these exercises are quite hard. The successful solution of a challenging problem gives a much better understanding of the powers and limitations of the theory than any number of trivial ones. Remember that mathematics is not a spectator sport!

PART I

The Algebraic Background

1

Groups

It is likely that the reader has already met the concept of a group. It was Galois who first understood the imporance of groups in the study of the roots of a polynomial equation; since then, group theory has blossomed, and developed as a subject in its own right. In this chapter we simply develop those parts of the theory which we shall need later; one of the main purposes is to explain the notation and terminology that we shall use.

1.1 Groups

Suppose that S is a set. A *law of composition* $\circ$ on S is a mapping from the Cartesian product $S \times S$ into S; that is, for each ordered pair (s_1, s_2) of elements of S there is defined an element $s_1 \circ s_2$ of S.

A group G is a non-empty set with a law of composition $\circ : G \times G \to G$ with the following properties:

(i) $g_1 \circ (g_2 \circ g_3) = (g_1 \circ g_2) \circ g_3$ for all g_1, g_2, g_3 in G – that is, composition is associative;

(ii) there is an element e in G (the *unit* or *neutral* element) such that $e \circ g = g \circ e = g$ for each g in G;

(iii) to each g in G there corresponds an element g^{-1} (the *inverse* of g) such that $g \circ g^{-1} = g^{-1} \circ g = e$.

Exercise

1.1   Suppose that G is a group. Show that the identity element e is unique, and that for each $g \in G$ the inverse element g^{-1} is also unique.

3

Two elements g and h of a group *commute* if $g \circ h = h \circ g$. The *commutator* $[g,h]$ of g and h is the element $g^{-1} \circ h^{-1} \circ g \circ h$; thus g and h commute if and only if $[g,h] = e$. A subset A of a group G is said to be *commutative*, or *abelian*, if and only if any two elements of A commute.

The notation that is used for the law of composition varies from situation to situation. Frequently, there is no symbol, and elements are simply placed side by side: $g \circ h = gh$. When G is abelian, it often happens that the law is denoted by $g \circ h = g + h$, the identity element is denoted by 0 and the inverse of an element g is denoted by $-g$.

Let us give some examples of groups. The integers $\mathbb{Z}$ (positive, zero and negative) form an abelian group under addition, with identity element 0, but the non-zero elements do not form a group under multiplication (2 has no multiplicative inverse in $\mathbb{Z}$). The non-zero complex numbers $\mathbb{C}^*$ form an abelian group under multiplication, with identity element 1.

If S is a non-empty set, a mapping σ from S to S is called a *permutation* of S if it is a bijection: that is, if $\sigma(x) = \sigma(y)$ then $x = y$, and if $z \in S$ there exists w in S for which $\sigma(w) = z$. The set Σ_S of all permutations of S is a group under the natural composition of mappings. It is not abelian if S has more than two elements. If $S = (1, \ldots, n)$, we write Σ_n for Σ_S. We shall consider S_n in more detail in Sections 1.3 and 1.4.

A subset H of a group G is a *subgroup* of G if it is a group under the law of composition defined on G; that is, if h_1 and h_2 are elements of H then so are $h_1 \circ h_2$ and h_1^{-1}. If G is a group, $\{e\}$ and G are subgroups; these are the *trivial* subgroups of G. If $n \in \mathbb{Z}$ the set $n\mathbb{Z} = \{nm : m \in \mathbb{Z}\}$ is a subgroup of $\mathbb{Z}$. The set $\mathbb{T} = \{z : |z| = 1\}$ is a subgroup of the multiplicative group $\mathbb{C}^*$, and if $n > 0$ the set $R_n = \{e^{2\pi i k/n} : 0 \le k < n\}$ of nth roots of unity is a subgroup of $\mathbb{T}$.

A group is *cyclic* if there is an element $g \in G$ such that every element of G is the composition of finitely many copies of g or or finitely many copies of g^{-1}.

A group is a *finite group* if it has finitely many elements. The *order* of a finite group is the number of its elements, and its *exponent* $e(G)$ is the smallest positive integer n such that $g^n = e$ for all $g \in G$.

Exercises

1.2   Show that if H is a subgroup of $\mathbb{Z}$, then H is cyclic.

1.3   Show that a subgroup F of a cyclic group is cyclic.

If $\{G_\alpha\}_{\alpha \in A}$ is a family of groups, then the product $\prod_{\alpha \in A} G_\alpha$ is a group, when composition is defined by $(g \circ h)_\alpha = g_\alpha \circ h_\alpha$ for $\alpha \in A$.

The intersection of subgroups of a group G is a subgroup, and so if S is a subset of a group G, there is a smallest group containing S, the subgroup generated by S; this is denoted by $\langle S \rangle$. It consists of all finite products of elements of S and their inverses, and is called the subgroup *generated by S*. For example, if G is a group, the *derived group* $\delta(G)$ is the subgroup generated by the set of all commutators $[g,h]$ in G. In the case where S is a singleton $\{a\}$, we write $\langle a \rangle$ for $\langle S \rangle$; $\langle a \rangle$ is then an abelian group, the *cyclic* subgroup generated by a, and consists of $\{a^n : n > 0\}$, e and $\{a^{-n} : n > 0\}$, where a^n is $a \circ a \circ \cdots \circ a$ (n terms) and a^{-n} is $a^{-1} \circ a^{-1} \circ \cdots \circ a^{-1}$ (n terms).

Suppose that A is a subset of a group G. The *centralizer* $Z(A)$ is the set $\{g \in G : [g,h] = e,\text{for all } h \in A\}$ of all elements of G which commute with every element of A.

Exercise

1.4 Suppose that A, A_1 and A_2 are subsets of a group G, and that $A_1 \subseteq A_2$. Show that

(i) $Z(A)$ is a subgroup of G
(ii) $Z(A_2) \subseteq Z(A_1)$
(iii) $A \subseteq Z(Z(A))$
(iv) $Z(A) = Z(Z(Z(A)))$
(v) $A \subset Z(A)$ if and only if A is abelian.

The group $Z(G)$ is called the *centre* of G; it is an abelian subgroup of G.

If G is a *finite group* (a group with finitely many elements) then the *order* of G is the number $|G|$ of elements of G. (If G is infinite, its order is ∞.) If $a \in G$ then the *order* of a is the order of $\langle a \rangle$. If a has finite order, the order of a is the least positive integer n such that $a^n = e$.

A mapping ϕ from a group G to a group H is a *homomorphism* if $\phi(g_1 \circ g_2) = \phi(g_1) \circ \phi(g_2)$ for all g_1 and g_2 in G. A homomorphism which is injective is called a *monomorphism*, one which is surjective is called an *epimorphism* and one which is both is called an *isomorphism*. If there is an isomorphism of a group G onto a group H, we say that G and H are *isomorphic* and write $G \cong H$. An isomorphism from a group onto itself is called an *automorphism*. For example, if $k \in G$, we set $g^k = k^{-1} \circ g \circ k$, for each $g \in G$. Then the mapping $g \to g^k$ is an automorphism of G (*conjugation by k*), an *inner automorphism* of G.

If A is a subset of a group G, and $k \in G$, we set $A^k = \{a^k : a \in A\}$. Two subsets A and B of a group G are *conjugate* if there exists $h \in G$ such that $B = A^h$; conjugacy is an equivalence relation on the subsets of G; we denote the equivalence class to which A belongs by $conj(A)$. If $\{g\}$ is a singleton, we write $conj(g)$; $conj(g)$ is called a *conjugacy class*. A set A is *self-conjugate* if $conj(A) = \{A\}$. Thus a group is abelian if and only if every singleton is self-conjugate.

A subgroup H of a group G is a *normal* subgroup if it is self-conjugate; if so, we write $H \triangleleft G$. Thus a subgroup H of G is normal if and only if it is the union of conjugacy classes. Every subgroup of an abelian group is normal. If g, h, k are elements of a group G then $[g, h]^k = [g^k, h^k]$, so that $\delta(G) \triangleleft G$.

Suppose that A is a non-empty subset of a group G. The *normalizer* $N(A) = N_G(A)$ is the set $N(A) = \{g \in G : a^g \in A \text{ for all } a \in A\}$. $N(A)$ is a subgroup of G; $Z(A) \subseteq N(A)$. We write $N(a)$ for $N(\{a\})$ and $Z(a)$ for $Z(\{a\})$; then $Z(a) = N(a)$ and $Z(A) = \bigcap_{a \in A} Z(a)$.

If H is a subgroup of G then $H \subseteq N(H)$ and $H \triangleleft N(H)$; if K is a subgroup of G containing H then $H \triangleleft K$ if and only if $K \subseteq N(H)$; thus $H \triangleleft G$ if and only if $N(H) = G$.

If ϕ is a homomorphism from a group G to a group H, its *image* $\phi(G)$ is a subgroup of H, and its *kernel* $\phi^{-1}(e)$ is a subgroup of G.

Exercise

1.5   Show that if ϕ is a homomorphism of a group G into a group H then its kernel is a normal subgroup of G.

If G is a group and $g \in G$, let $r_g(h) = h \circ g$ and $l_g(h) = g^{-1} \circ h$, for each $h \in G$. Then l_g and r_g are permutations of G, and the mappings $g \to r_g$ and $g \to l_g$ are monomorphisms of G into the permutation group Σ_G. If H is a subgroup of G then a set $r_g(H) = H \circ g$ is called a *right coset* and a set $l_g(H) = g^{-1} \circ H$ is called a *left coset*. (Terminology here varies, but this seems preferable.)

Proposition 1.1 *Let H be a subgroup of a group G. If g_1 and g_2 are elements of G then either $g_1^{-1} \circ g_2 \in H$, in which case $r_{g_1}(H) = r_{g_2}(H)$, or $g_1^{-1} \circ g_2 \notin H$, in which case $r_{g_1}(H)$ and $r_{g_2}(H)$ are disjoint. Thus the distinct right cosets form a partition of G. Similarly for left cosets.*

Proof   Simple verification.   □

The set of right cosets H in G is denoted by G/H. The number $|G/H|$ of right cosets H in G is called the *index* of H in G.

Exercise

1.6   Suppose that A is a non-empty subset of a group G and that $g \in G$. Let $\psi_A(g) = A^g$. Show that ψ_A maps G onto $conj(A)$, and $\psi_A(g') = \psi_A(g)$

if and only if $g' \in N(A)g$. Thus if G is a finite group then $|conj(A)| = |G/N(A)|$, so that $|conj(A)|.|N(A)| = |G|$.

Proposition 1.2 (Lagrange's theorem) *If H is a subgroup of a finite group G then $|G| = |G/H|.|H|$.*

Proof For if $H \circ g \in G/H$, $H \circ g = r_g(H)$, so $|Hg| = |H|$; the cosets of H have the same number of elements as H.   □

Corollary 1.3 *(i) If a is an element of a finite group G and a has order k, then k divides $|G|$.*
  (ii) If G is a cyclic group of order n, and $a \in G$, then $a^n = e$.

The left and right cosets need not be the same.

Proposition 1.4 *Suppose that H is a subgroup of a group G. Then H is normal in G if and only if the left cosets and right cosets of H are the same.*

Proof Suppose that H is a normal subgroup of G. If $g \in G$ and $h \in H$, let $h' = h^g$ and $h'' = h^{g^{-1}}$. Then $h \circ g = g \circ h'$ and $g \circ h = h'' \circ g$, from which it follows that $g \circ H = H \circ g$.
  Conversely, suppose that the left cosets and right cosets of H are the same. If $g \in G$ then $C = H \circ g$ is a right coset, and is therefore a left coset. Since $g \in C$, $C = g \circ H$, so that $g \circ H = H \circ g$, and $h^g \in H$ for each $h \in H$. Since this holds for each $g \in G$, H is normal in G.   □

Exercise

1.7   Show that if G is a finite group and H is a subgroup of G with index 2 in G, then H is a normal subgroup of G.

Theorem 1.5 *Suppose that $H \triangleleft G$ and that $C_1 = H \circ g_1$ and $C_2 = H \circ g_2$ are two cosets of H in G. Then $C_1 \circ C_2$ is again a coset of H in G, and under this law of composition, G/H is a group (the quotient group) with H as neutral element, and the quotient mapping $q : G \to G/H$ (which sends g to $H \circ g$) is an epimorphism of G onto G/H, with kernel H.*

Proof

$$C_1 \circ C_2 = \{h \circ g_1 \circ k \circ g_2 : h, k \in H\}$$
$$= \{h \circ k^{g_1^{-1}} \circ (g_1 \circ g_2) : h, k \in H\}$$
$$= \{h \circ (g_1 \circ g_2) : h \in H\} = H \circ (g_1 \circ g_2),$$

from which the result follows easily.   □

Exercise

1.8   Show that if G is a group, then $G/\delta(G)$ is abelian, and if $H \lhd G$ then
      G/H is abelian if and only if $\delta(G) \subseteq H$.

Theorem 1.6 (The first isomorphism theorem) *Suppose that ϕ is a homomorphism from a group G_1 into a group G_2, with kernel H. Then $H \lhd G_1$, and
there is an isomorphism $\tilde{\phi}$ from G/H onto $\phi(G_1)$ such that $\phi = i.\tilde{\phi}.q$, where
$q : G \rightarrow G/H$ is the quotient mapping and $i : \phi(G_1) \rightarrow G_2$ is the inclusion
mapping (which is of course a monomorphism).*

Proof   If $C = H \circ g \in G_1/H$, let $\tilde{\phi}(C) = \phi(g)$. This does not depend on the
choice of g in C, $\tilde{\phi}$ is a homomorphism of G/H onto $\phi(G_1)$, and $\tilde{\phi}(C) = e$ if
and only if $C = H$. Then $\phi = q.\tilde{\phi}.i$ is an isomorphism of G_1/H onto $\phi(G_1)$:

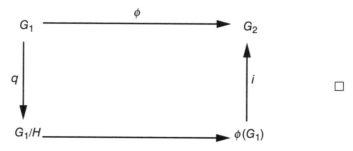

As an example, if $n > 0$ let $\phi(k) = e^{2\pi i k/n}$. Then ϕ is a homomorphism
of $(\mathbb{N}, +)$ into $(\mathbb{T}, \times)$, with kernel $n\mathbb{Z}$ and image $\mathbb{T}_n$, the set of nth roots of
unity. We write $(\mathbb{Z}_n, +)$ for the group $(\mathbb{Z}/n\mathbb{Z}, +)$. Thus $(\mathbb{Z}_n, +)$ is isomorphic
to $(\mathbb{T}_n, \times)$: $\mathbb{Z}_n$ is the set of equivalence classes of $\mathbb{Z}$ (mod n).
   Here is an application of the first isomorphism theorem.

Theorem 1.7 *Suppose that G is a group, that $H \lhd G$ and that A is a subgroup
of G.*

(i) $H \cap A \lhd A$, and $A/(H \cap A) \cong \langle H, A \rangle/H$.
*(ii) If $H \subseteq A$ and $A \lhd G$, then $H \lhd A$, $A/H \lhd G/H$ and $(G/H)/(A/H) \cong$
G/A.*

Proof   (i) If $h \in H \cap A$ and $a \in A$ then $h^a \in H \cap A$, so that $H \cap A \lhd A$.
If $j : A \rightarrow G$ is the inclusion mapping and $q : G \rightarrow G/H$ is the quotient
mapping then $q \circ j$ is a homomorphism from A into G/H with kernel $H \cap A$
and image $\langle H, A \rangle/H$, so that the result follows from the first isomorphism
theorem.
   (ii) Certainly $H \lhd A$, by (i). If C is a right coset of H in G, let $A \circ C =
\{a \circ c : a \in A, c \in C\}$; $A \circ C$ is a right coset of A in G. Let $\theta(C) = A \circ C$.

Then $\theta : G/H \to G/A$ is an epimorphism with kernel A/H, so that the result also follows from the first isomorphism theorem. $\qquad\square$

Exercises

1.9 Suppose that G has exactly one subgroup H of order k. Show that H is a normal subgroup of G.

1.10 Suppose that H is a normal subgroup of G and that K is a normal subgroup of H. Is K necessarily a normal subgroup of G?

1.11 Show that a group G is generated by each of its elements (other than the unit element) if and only if G is a finite cyclic group of prime order.

1.12 Describe the elements of $\mathbb{Z}_n$ which generate $\mathbb{Z}_n$ (for any positive integer n).

1.13 Give an example of a non-abelian group of order 8 all of whose subgroups are normal.

From now on, if g and h are elements of a group G, we shall write gh for $g \circ h$ (unless G is an abelian group, written additively).

1.2 Finite Abelian Groups

A finite cyclic group such as $\mathbb{Z}_n$ or $\mathbb{T}_n$ is a very simple example of a finite abelian group. Any finite abelian group is isomorphic to a finite product of cyclic groups.

Theorem 1.8 *Suppose that $(G,+)$ is a finite abelian group. G is isomorphic to a product of cyclic groups:*

$$G \cong \mathbb{Z}_{d_1} \times \cdots \times \mathbb{Z}_{d_s}.$$

Further, the isomorphism can be chosen so that $d_j | d_k$ for $1 \leqslant j < k \leqslant s$. The number s is characterized by the property that G is generated by s elements, but it is not generated by $s - 1$ elements.

Proof We prove this by induction on $|G|$. Suppose that the result is true for all abelian groups of order less than n, and that $|G| = n$.

There exists an integer s such that G is generated by s elements, but is not generated by fewer than s elements. Let m be the least positive number such that there exists a set $\{g_1, \ldots, g_s\}$ of generators and a relation

$$mg_1 + a_2 g_2 + \cdots + a_s g_s = 0$$

(with $a_2, \ldots, a_s$ in $\mathbb{Z}$). Note that $m > 1$, since otherwise G would be generated by $\{g_2, \ldots, g_s\}$. We can write $a_i = mq_i + r_i$ with $0 \leqslant r_i < m$, for $2 \leqslant i \leqslant s$. Then if $h_1 = g_1 + q_2 g_2 + \cdots + q_s g_s$, G is generated by $\{h_1, g_2, \ldots, g_s\}$ and

$$mh_1 + r_2 g_2 + \cdots + r_s g_s = 0.$$

The minimality of m implies that $r_2 = r_3 = \cdots = r_s = 0$, and so $mh_1 = 0$. We now claim that G is isomorphic to $\langle h_1 \rangle \times \langle g_2, \ldots, g_s \rangle$. If $(a, b) \in \langle h_1 \rangle \times \langle g_2, \ldots, g_s \rangle$, let $\theta(a, b) = a + b$. The map θ is a homomorphism of $\langle h_1 \rangle \times \langle g_2, \ldots, g_s \rangle$ into G. It is an epimorphism, since $\{h_1, g_2, \ldots, g_s\}$ generates G. If (a, b) is in the kernel of θ, $a + b = 0$. Writing $a = j_1 h_1, b = j_2 g_2 + \cdots + j_r g_s$, with $0 \leqslant j_1 < m$, we have

$$j_1 h_1 + j_2 g_2 + \cdots + j_s g_s = 0.$$

It follows from the minimality of m that $j_1 = 0$, and so $a = b = 0$. Thus θ is an isomorphism.

We now apply the inductive hypothesis to $\langle g_2, \ldots, g_s \rangle$, which is clearly generated by $s - 1$ elements, but not by $s - 2$ elements: the subgroup $\langle g_2, \ldots, g_s \rangle$ is isomorphic to

$$\mathbb{Z}_{d_2} \times \cdots \times \mathbb{Z}_{d_s},$$

with $d_j | d_k$ for $2 \leqslant j < k \leqslant s$. Consequently $G \cong \mathbb{Z}_m \times \mathbb{Z}_{d_2} \times \cdots \times \mathbb{Z}_{d_s}$. Let $h_1, \ldots, h_s$ be the corresponding generators in G. It follows from the minimality of m that $m \leqslant d_2$. Let $d_2 = e_2 m + f_2$, where $0 \leqslant f_2 < m$, and let $h_1' = h_1 + e_2 h_2$. Then G is generated by $\{h_1', h_2, \ldots, h_s\}$. As

$$mh_1' + f_2 h_2 = 0$$

it follows that $f_2 = 0$, and so $m | d_2$. This completes the proof. $\qquad \square$

Corollary 1.9 *If G is a finite abelian group, there exists $g \in G$ whose order is its exponent $e(G)$.*

We can extend this theorem. We begin with a lemma.

Lemma 1.10 *If G is a cyclic group of order jk, where j and k are coprime (that is, if n divides both j and k, then $n = 1$) then G is isomorphic to a product $J \times K$, where J is a cyclic group of order j and K is a cyclic group of order k.*

Proof Let a be a generator of G, and let J and K be cyclic groups of orders j and k, with generators b and c, respectively. If $h = (rb, sc) \in J \times K$, let $\phi(h) = (kr + js)a$. Then $\phi : J \times K \to G$ is a homomorphism. Since j and k

are coprime, there exist r and s such that $kr + js = 1$. It therefore follows that $a \in \phi(J \times K)$, and so ϕ is surjective. But $J \times K$ has the same order as G, and so ϕ is an isomorphism. □

Theorem 1.11 *If $(G, +)$ is a finite abelian group, then G is isomorphic to a product of cyclic groups of prime power order.*

Proof It follows by induction that a cyclic abelian group is the product of cyclic groups of prime power order, and the result then follows from Theorem 1.8. □

Exercises

1.14 Suppose that G is a finite abelian group for which every element other than the identity has order k. Show that k is a prime number, and that G is isomorphic to the product of cyclic groups, each of order k.

1.15 Suppose that a and b are positive integers with highest common factor d. Show that

$$\mathbb{Z}_a \times \mathbb{Z}_b \cong \mathbb{Z}_d \times \mathbb{Z}_{ab/d}.$$

1.16 Suppose that G is an abelian group. Show that the set T of elements of finite order is a subgroup of G and that every element of G/T, except the identity, is of infinite order.

1.17 Suppose that G is a finitely generated abelian group every element of which, except the identity, has infinite order. Show that $G \cong \mathbb{Z}^s$, where s is defined by the property that G is generated by s elements, but is not generated by $s - 1$ elements.

1.18 Suppose that G is a finitely generated abelian group. Show that $G \cong \mathbb{Z}^s \times T$, where T is a finite group.

1.3 Finite Permutation Groups

We now describe the terminology that we use concerning permutation groups, and establish some basic facts about them.

Suppose that G is a subgroup of Σ_S. Set $x \sim_G y$ if there is $\sigma \in G$ such that $y = \sigma(x)$. This is an equivalence relation on S. The equivalence classes are called the *orbits* O_G of G; we denote the orbit to which x belongs by $O_G(x)$. Let $stab_G(x) = \{g \in G : g(x) = x\}$; $stab_G(x)$ is the *stabilizer* of x in G.

$stab_G(x)$ is a subgroup of G. If $g \in G$, let $\phi(g) = g(x)$: then ϕ maps G onto $O_G(x)$, and $\phi(g') = \phi(g)$ if and only if g' is in the left coset $g.stab_G(x)$. Thus we have the following.

Proposition 1.12 *Suppose that S is a finite set, that G is a subgroup of Σ_S and that $x \in S$. Then $|G| = |stab_G(x)|.|O_G(x)|$.*

G is said to act *transitively* on S if S is an orbit.

Corollary 1.13 *Suppose that S is a finite set, that $|S| = p$, a prime number, and that G is a transitive subgroup of Σ_S. Then p divides $|G|$.*

Suppose that $\sigma \in \Sigma_S$. Let $\langle \sigma \rangle$ be the cyclic subgroup of Σ_S generated by σ. We then write $x \sim_\sigma y$ for $x \sim_{\langle \sigma \rangle} y$ and write $O_\sigma(x)$ for $O_{\langle \sigma \rangle}(x)$.

If C is an orbit, let $\sigma_{|C}$ be the restriction of σ to C. Then $\langle \sigma_{|C} \rangle$ is a cyclic subgroup of Σ_S, which acts transitively on C and leaves the elements of $S \setminus C$ fixed.

Suppose that S is finite. We label S as $\{0, 1, \ldots, n-1\}$, where $n = |S|$, and refer to Σ_S as Σ_n. If $\sigma \in \Sigma_n$ and $C = O_\sigma(a)$ is one of the orbits of σ we write $\sigma_{|C}$ as $(a_1, \ldots, a_j)$, where $j = |C|$, $a_1 = a$, $a_i = \sigma(a_{i-1})$ for $2 \leq i \leq j$ and $a_1 = \sigma(a_j)$. $\sigma_{|C}$ is called a *j-cycle*, or *cycle of type j*. Thus we can write σ as a product of disjoint cycles: for example, in Σ_{12} we have the permutation $(1, 7)(2, 9)(3, 5, 6)(8, 10, 4, 12)$, the product of two 2-cycles, one 3-cycle and one 4-cycle. (A 2-cycle is called a *transposition*.) The *cycle type* of σ is then a list of the cycle types of its constituent disjoint cycles. Thus the example above has cycle type $2^2.3.4$.

We can also consider the product of cycles which are not disjoint. For example:

$$(a_1, a_2, \ldots, a_j) = (a_1, a_j)(a_1, a_{j-1}) \ldots (a_3, a_1)(a_2, a_1)$$

is the product of $j - 1$ transpositions, and so it follows that any permutation can be written as a product of transpositions.

Proposition 1.14 *Two permutations in Σ_n are conjugate in Σ_n if and only if they have the same cycle type.*

Proof Suppose that $\sigma \in \Sigma_n$ and that $\tau = (a, b)$ is a transposition. Then $\sigma^\tau = \tau^{-1}\sigma\tau$ is obtained by replacing a by b and b by a whenever they occur in the expression for σ.   □

Proposition 1.15 *Suppose that S is finite, and that G is a transitive subgroup of Σ_S. If H is a normal subgroup of G, then the orbits of H are of the same size.*

Proof Suppose that $x, y \in S$ and that $x' \in O_H(x)$. There exists $\tau \in G$ such that $\tau(x) = y$ and $\sigma \in H$ such that $\sigma(x) = x'$. Then $\tau(x') = \tau\sigma(x) = \tau\sigma\tau^{-1}(y) \in O_H(y)$, since $\tau\sigma\tau^{-1} \in H$, so that τ is an injective mapping of $O_H(x)$ into $O_H(y)$. Similarly, τ^{-1} is an injective mapping of $O_H(y)$ into $O_H(x)$, and so $|O_H(x)| = |O_H(y)|$. $\qquad\square$

Corollary 1.16 *Suppose that* $|S| = p$, *a prime number, and that* H *is a proper normal subgroup of a transitive subgroup* G *of* Σ_S. *Then* H *is transitive.*

If we label S as $\{0, 1, \ldots, n-1\}$ and let $\alpha(j) = j + i \pmod{n}$, then $Z_n = \{\alpha^k : 0 \le k < n\}$ is a transitive cyclic subgroup of Σ_n of order n, which represents addition in $\mathbb{Z}_n$. Z_n is generated by α.

If we set $\tau(j) = -j$, then $\{e, \tau\}$ is a subgroup of Σ_n of order 2. If $n = 2k$ is even, it is the product of $k-1$ disjoint transpositions, and has two fixed points, 0 and k, and so has cycle type 2^{k-1}. If $n = 2k+1$ is odd, it is the product of k disjoint transpositions, and has one fixed point 0, and so has cycle type 2^k. Since $\alpha\tau = \tau\alpha^{-1}$, the subgroup generated by σ and τ is non-abelian for $n > 3$. It is known as the *dihedral group* D_n; it is the group of symmetries (rotations and reflections) of a regular polygon with n vertices. Note that every element of D_n can be written as τ, α^b or $\tau\alpha^b$, and that $Z_n \lhd D_n$, but that $\{0, \tau\} \lhd D_n$ only if $n = 2$.

Let $\mathbb{Z}_p^* = \mathbb{Z}_p \setminus \{0\}$, where p is a prime number. Suppose that $r, s, t \in \mathbb{Z}_p^*$ and that $s \ne t$. Then $r(s - t) \ne 0 \pmod{p}$, so that $rs \ne rt$. Thus the map $\mu_r : s \to rs$ is an injective map from $\mathbb{Z}_p^*$ to itself, and is therefore a bijection. In particular, there exists s so that $rs = 1$. Since $rs = sr$, $(rs)t = r(st)$ and $1r = r$, it follows that $\mathbb{Z}_p^*$ is an abelian group under multiplication, which we denote by M_p.

Proposition 1.17 *If* p *is a prime number then* $(\mathbb{Z}_p^*, M_p)$ *is a cyclic group.*

Proof Let λ be the exponent of $(\mathbb{Z}_p^*, M_p)$. λ is a divisor of $p - 1$. Suppose that $\lambda < p - 1$; $x^\lambda = 1$ for $1 \le x \le \lambda$, so that $x^\lambda - 1 = \prod_{j=1}^{\lambda}(x - j)$. But then $(\lambda + 1)^\lambda - 1 = \lambda! \ne 0 \pmod{p}$, giving a contradiction. Thus $\lambda = p - 1$ and $(\mathbb{Z}_p^*, M_p)$ is cyclic. $\qquad\square$

We shall extend this result later, when we consider fields, in Section 2.5. We can deduce the following result from elementary number theory.

Corollary 1.18 (Wilson's theorem) *If* p *is an odd prime number then* $(p - 1)! = -1 \pmod{p}$.

Proof We consider the numbers $\{1, 2, \ldots, p-1\}$ as the elements of $(\mathbb{Z}_p^*, M_p)$. $1^{-1} = 1$ and $(p - 1)^{-1} = p - 1$. Otherwise $a^{-1} \ne a \pmod{p}$, so that the set of numbers from 2 to $p-2$ is the union of $(p-3)/2$ disjoint pairs $\{a, a^{-1}\}$, and so $(p-2)! \equiv 1 \pmod{p}$. Since $p-1 \equiv -1 \pmod{p}$, the result follows. $\square$

We have considered M_p acting on $\mathbb{Z}_p^*$; we extend M_p to act on $\mathbb{Z}_p$ in the obvious way, by setting $\mu_r(0) = 0$. We also consider the *affine subgroup* W_p of Σ_p generated by Z_p and M_p. Since $\mu_r \alpha^k(x) = rx + rk = \alpha^{rk} \mu_r(x)$, each element of W_p can be written uniquely as $\sigma_{r,k} = \alpha^k \mu_r$, where $k \in \mathbb{Z}_p$ and $r \in \mathbb{Z}_p^*$. Thus $|W_p| = p(p-1)$. Further, the map $P : \sigma_{r,k} \to \mu_r$ is a homomorphism of W_p onto M_p with kernel Z_p, so that $Z_p \triangleleft W_p$.

If $\sigma \in \Sigma_n$, its *signature* ϵ_σ is defined to be

$$\frac{\prod_{i<j}(\sigma(j) - \sigma(i))}{\prod_{i<j}(j - i)}.$$

Since the terms in the numerator are, except possibly for sign changes, the same as the terms in the denominator, $\epsilon_\sigma = \pm 1$. Since

$$\epsilon_{\sigma\tau} = \frac{\prod_{i<j}(\sigma\tau(j) - \sigma\tau(i))}{\prod_{i<j}(\tau(j) - \tau(i))} \cdot \frac{(\prod_{i<j}(\tau(j) - \tau(i))}{\prod_{i<j}(j - i)} = \epsilon_\sigma . \epsilon_\tau,$$

ϵ is a homomorphism of Σ_n into $R_2 = (\{1, -1\}, \times)$. Since $\sigma_{(12)} = -1$, it is surjective. Its kernel A_n is therefore a normal subgroup of Σ_n of index 2. It is called the *alternating group of degree n*.

Let us consider Σ_n and A_n for $n = 2, 3, 4$ and 5. Σ_2 has two elements, and is isomorphic to $\mathbb{T}_2$, and A_2 only has one element, namely the identity. Σ_3 has six elements, the identity e, three 2-cycles and two 3-cycles. It is not commutative. A_3 has three elements, e and two 3-cycles. It is abelian, and is isomorphic to $\mathbb{T}_3$.

Let us tabulate Σ_4 and Σ_5.

σ_4

Cycle type	Signature	Typical element	How many?
e	1	e	1
2^2	1	(01)(23)	3
3	1	(012)	8
2	-1	(01)	6
4	-1	(0123)	6

The first three rows represent the elements of A_4.

σ_5

Cycle type	Signature	Typical element	How many?	Fixing 0
e	1	e	1	1
5	1	(01234)	24	0
2^2	1	(12)(34)	15	3
3	1	(123)	20	8
2	-1	(12)	10	6
2.3	-1	(01)(234)	20	0
4	-1	(1234)	30	6

The first four rows represent the elements of A_5. The last column represents a subgroup isomorphic to Σ_4.

There is an epimorphism of Σ_4 onto Σ_3, which is best expressed geometrically. Let T be the tetrahedron in $\mathbb{R}^3$ with vertices $a_1 = (1,1,1)$, $a_2 = (1, -1, -1)$, $a_3 = (-1, 1, -1)$ and $a_4 = (-1, -1, 1)$. T has six edges, with midpoints $(\pm 1, 0, 0)$, $(0, \pm 1, 0)$ and $(0, 0, \pm 1)$, so that the lines joining the midpoints of opposite edges are the axes $X = \{(x, 0, 0) : x \in \mathbb{R}\}$, $Y = \{(0, y, 0) : y \in \mathbb{R}\}$ and $Z = \{(0, 0, z) : z \in \mathbb{R}\}$. Then a permutation σ of $\{a_1, a_2, a_3, a_4\}$ produces a permutation $\phi(\sigma)$ of $\{X, Y, Z\}$, and ϕ is clearly a homomorphism. If σ is the transposition (a_1, a_2), $\phi(\sigma)$ is the transposition (Y, Z), and similarly for other transpositions, so that ϕ is an epimorphism. Its kernel is the four-element subgroup

$$N = \{e, (a_1, a_2)(a_3, a_4), (a_1, a_3)(a_2, a_4), (a_1, a_4)(a_2, a_3)\}.$$

This is usually known by its German name of *Vierergruppe*.

Let $G_0 = \{e\}$, $G_1 = \{e, (1, 2)\}$, $G_2 = N$, $G_3 = A_4$ and $G_4 = \Sigma_4$. Then G_i is a normal subgroup of G_{i+1} for $0 \leq i \leq 3$, and

$$G_1/G_0 \sim \mathbb{Z}_2, \ G_2/G_1 \sim \mathbb{Z}_2, G_3/G_2 \sim \mathbb{Z}_3 \ and \ G_4/G_3 \sim \mathbb{Z}_2'.$$

(Note however that G_1 is not a normal subgroup of G_3.) This lies behind the solution of quartic equations, and we shall study this phenomenon in general in the next section.

The situation is completely different when $n \geq 5$. A group G is *simple* if it has no non-trivial normal subgroups: $\{e\}$ and G are the only normal subgroups of G.

Theorem 1.19 A_n *is simple, for* $n \geq 5$.

Proof Let B_n be the set of products of two disjoint transpositions, and let T_n be the set of 3-cycles in A_n. Since $(a_1a_3)(a_1a_2) = (a_1a_2a_3)$, $A_n = \langle B_n \cup T_n \rangle$, and since $(a_1a_2a_3) = (a_4a_5)(a_1,a_3)(a_4a_5)(a_1a_2)$, $T_n \subseteq \langle B_n \rangle$.

It is therefore sufficient to show that if N is a normal subgroup of A_n other than $\{e\}$, then $B_n \subseteq N$. We prove this by induction on n, for $n \geq 4$. The result is true for $n = 4$, since the Vierergruppe and A_4 both contain B_4. Suppose that $n > 4$ and that the result holds for $n - 1$. Let $F_j = \{\sigma \in A_n : \sigma(j) = j\}$. Each F_j is a subgroup of A_n isomorphic to A_{n-1}, and therefore contains $B_n \cap F_j$.

We shall show that there exists j such that $N \cap F_j \neq \{e\}$. Suppose that $\tau \in N$ and that $\tau \neq e$. We consider two possibilities. First, τ is a product of two or more disjoint transpositions, $(a_1a_2)(a_3a_4)\rho$, say. Let $\sigma = (a_1a_2a_3)$. Then $\tau' = \tau^\sigma = (a_1a_3)(a_2a_4) \in N$, and $\tau\tau' = (a_1a_4)(a_2a_3) \in N$, so that there exists j such that $N \cap F_j \neq \{e\}$ for some j. Second, in its cycle representation τ contains a cycle of length greater than 2, so that $\tau = (a_1a_2 \ldots a_k)\rho$, and τ is not in any F_j. Let j be different from a_1, a_2 and a_3, let $\sigma = (a_1, j)(a_2, \tau(j))$ and let $\tau' = \tau^\sigma$, so that $\tau' = (j, \tau(j), a_3 \ldots)\rho' \in N$. Then $\tau(j) = \tau'(j)$, while $\tau'(\tau(j)) = a_3 = \tau\tau(a_1) \neq \tau(\tau(j))$, so that $\tau^{-1}\tau'$ is an element of $N \cap F_j$ different from $\{e\}$.

Thus there exists j such that $N \cap F_j \neq \{e\}$. But $N \cap F_j$ is a normal subgroup of $A_n \cap F_j$, which is isomorphic to A_{n-1}, and so $B_n \cap F_j \subseteq N \cap F_j$. But if $\tau = (ij)(kl)$ is an element of B_n which moves j, and m is different from j, let $\sigma = (jm)(kl)$ and $\tau' = \tau^\sigma = (im)(kl) \in B_n \cap F_j \subseteq N \cap F_j$, so that $\tau \in N$. Thus $B_n \subseteq N$, and so $N = A_n$. $\qquad\square$

Here is another result that we shall need later.

Theorem 1.20 *Suppose that G is a transitive subgroup of Σ_n. Let $i \sim j$ if either $i = j$ or $(i, j) \in G$. Then $\sim$ is an equivalence relation on $\{0, \ldots, n-1\}$ which partitions $\{0, \ldots, n-1\}$ into equivalence classes of equal size. In particular, if n is a prime number and G contains a transposition, then $G = \Sigma_n$.*

Proof Certainly $i \sim i$ and if $i \sim j$ then $j \sim i$. Suppose that $i \sim j$ and $j \sim k$. Then since $(i, j)(j, k)(i, j) = (i, k)$, $\sim$ is an equivalence relation. Consequently, $\{1, \ldots, n\}$ is partitioned into equivalence classes. Suppose that O_i and O_j are two distinct equivalence classes. Since G is transitive, there exists $g \in G$ such that $g(i) = j$. Suppose that $k \in O_i$ and that $l = g(k)$. Then $g(i, k)g^{-1} = (j, l)$ and $g(i, k)g^{-1} \in G$, so that $l \in O_j$ and $g(O_i) = O_j$. Thus the equivalence classes have the same size. If n is a prime, there is only one orbit; G contains all transpositions, and so $G = \Sigma_n$. $\qquad\square$

Exercises

1.19 By considering conjugacy classes, show directly that A_5 is simple.

1.20 Show that the group of rotations of the cube has 24 elements. By considering its four diagonals, show that it is isomorphic to Σ_4. By considering the three pairs of opposite faces, show that there is an epimorphism of Σ_4 onto Σ_3.

1.21 Show that the group of rotations of the dodecahedron has 60 elements. Using the fact that five cubes can be inscribed in a dodecahedron, or otherwise, show that it is isomorphic to A_5.

1.4 Group Series

Suppose that G is a group. A *group series* for G is a strictly decreasing finite sequence $(G_i)_{i=0}^r$ of subgroups of G, with $G_0 = G$ and $G_r = \{e\}$. Group series take many forms, and play a large part in the theory of groups.

A *normal series* (sometimes called a *subnormal series*) is a group series for which $G_i \lhd G_{i-1}$ for $1 \le i \le r$.

A group G is *soluble* if it has a normal series $(G_i)_{i=0}^r$ for which G_i/G_{i-1} is abelian, for $1 \le i \le r$. Soluble groups play a major role in Galois theory, and we shall consider soluble groups further in the next section.

A *derived series* is a group series $(G_i)_{i=0}^r$ for which $G_i = \delta(G_{i-1})$ for $1 \le i \le r$.

Proposition 1.21 *A group G is soluble if and only if it has a derived series.*

Proof It follows from the properties of δ that if $(G_i)_{i=0}^r$ is a derived series, then it is a normal series, and G_{i-1}/G_i is abelian for $1 \le i \le r$, so that G is soluble. Conversely, suppose that G is soluble, and that $(G_i)_{i=0}$ is a normal series with abelian quotients. Let $D_0 = G$ and let $D_i = \delta(D_{i-1})$, for $1 \le i \le r$. We prove by induction that $D_i \subseteq G_i$, for $1 \le i \le r$. It is certainly true for $i = 1$. Suppose that it is true for $i - 1$. Then $D_i = \delta(D_{i-1}) \subseteq \delta(G_{i-1}) \subseteq G_i$. It follows that there exists $s \le r$ such that $D_{s-1} \ne D_s = e$. □

An *invariant series* is a group series for which $G_i \lhd G$ for $0 \le i \le r - 1$. Clearly an invariant series is a normal series. Any inner automorphism of G leaves an invariant series unchanged.

Recall that if H is a subgroup of a group G then the centralizer $Z(H)$ is the set $\{g : [g,h] = e$ for h in $H\}$. A *central series* is an invariant series $(G_i)_{i=0}^r$ for which $G_{i-1}/G_i \subseteq Z(G/G_i)$ for $1 \le i \le r$. A group with a central series is called a *nilpotent* group. A nilpotent group is a soluble group, but a soluble group need not be nilpotent (consider Σ_3).

Proposition 1.22 *An invariant series is a central series if and only if* $[g, h] \in G_i$ *for every* $g \in G$ *and* $h \in G_{i-1}$, *for* $1 \le i \le r$.

Proof   The series is central if and only if $[gG_i, hG_i] = G_i$, for $g \in G$ and $h \in G_{i-1}$ for $1 \le i \le r$. This happens if and only if $[g, h] \in G_i$ for $g \in G$ and $h \in G_{i-1}$, for $1 \le i \le r$.   □

Exercise

1.22   A group series $(G_i)_{i=0}^r$ is an *upper central series* if $G_{i-1}/G_i = Z(G/G_i)$ for $1 \le i \le r$. Show that G possesses an upper central series if and only if G is nilpotent.

Nilpotent groups are important in group theory, but we shall not consider them further.

1.5 Soluble Groups

Recall that group G is *soluble* if there is a finite sequence of subgroups

$$\{e\} = G_n \subset G_{n-1} \subset \cdots \subset G_1 \subset G_0 = G$$

such that $G_j \triangleleft G_{j-1}$ and G_{j-1}/G_j is abelian for $1 \le j \le n$.

Thus $\{e\} \triangleleft A_3 \triangleleft \Sigma_3$, $A_3 \sim \mathbb{Z}_3$ and $\Sigma_3/A_3 \sim \mathbb{Z}_2$, so that Σ_3 is soluble. Similarly, $\{e\} \triangleleft N \triangleleft A_4 \triangleleft \Sigma_4$, and the quotients are abelian, so that Σ_4 is soluble. But the groups A_n for $n \ge 5$ are not soluble, since they are simple and not abelian.

Theorem 1.23 *If G is a finite group, G is soluble if and only if there is a finite sequence of subgroups*

$$\{e\} = G_n \subset G_{n-1} \subset \cdots \subset G_1 \subset G_0 = G$$

such that $G_j \triangleleft G_{j-1}$ *and* G_{j-1}/G_j *is cyclic of prime power order for* $1 \le j \le n$.

Proof   The condition is certainly sufficient. Suppose that G is soluble, and that

$$\{e\} = G_n \subset G_{n-1} \subset \cdots \subset G_1 \subset G_0 = G$$

is a sequence for which G_{j-1}/G_j is abelian for $1 \leq j \leq n$. Fix j. There exists a sequence $G_{j-1}/G_{j-1} \subset A_1 \cdots A_k = G_j/G_{j-1}$ such that A_l/A_{l-1} is cyclic of prime power order, for $1 \leq l \leq k$. Let $B_l = q^{-1}(A_l)$ (where $q : G_j \to G_{j-1}$ is the quotient map) for $1 \leq l \leq k$. By Theorem 1.7, $G_{j-1} \triangleleft B_1 \triangleleft \cdots \triangleleft B_k = G_j$. Thus if we do this for each j, we obtain a sequence for which all the quotients are cyclic of prime power order: the condition is necessary. $\qquad\square$

Theorem 1.24 (i) *If G is a soluble group and A is a subgroup of G then A is soluble.*

(ii) *Suppose that G is a group and $H \triangleleft G$. Then G is soluble if and only if H and G/H are soluble.*

Proof (i) Let $\{e\} = G_n \triangleleft G_{n-1} \triangleleft \cdots \triangleleft G_0 = G$, such that G_{i-1}/G_i is cyclic for $1 \leqslant i \leqslant n$. Let $A_i = A \cap G_i$. Then $A_i = A_{i-1} \cap G_i \triangleleft A_{i-1}$ and

$$A_{i-1}/A_i \cong G_i A_{i-1}/G_i$$

by the previous theorem. But $G_i A_{i-1}/G_i$ is a subgroup of the cyclic group G_{i-1}/G_i, so that either $A_{i-1} = A_i$ or A_{i-1}/A_i is cyclic.

(ii) First suppose that G is soluble, and that $\{e\} = G_n \triangleleft \cdots \triangleleft G_0 = G$, with G_{i-1}/G_i cyclic. H is soluble, by (i). Let $H_i = HG_i/H$. Now if $Hg_i \in H_i$ and $Hg_{i-1} \in H_{i-1}$:

$$(Hg_{i-1})^{-1} Hg_i Hg_{i-1} = g_{i-1}^{-1} Hg_i Hg_{i-1}$$
$$= Hg_{i-1}^{-1} g_i g_{i-1} \in H_i$$

so that $H_i \triangleleft H_{i-1}$.

Let $q : G \to G/H$ be the quotient mapping and let $q_i : H_{i-1} \to H_{i-1}/H_i$ denote the quotient mapping. Then $q(G_{i-1}) = H_{i-1}$, and $q_i q$ maps G_{i-1} onto H_{i-1}/H_i. An element g of G_{i-1} is in the kernel of $q_i q$ if and only if $Hg \in HG_i$: that is, if and only if $g \in HG_i$. The restriction of $q_i q$ to G_{i-1} therefore has kernel $HG_i \cap G_{i-1}$ so that by the first isomorphism theorem:

$$H_{i-1}/H_i \cong G_{i-1}/(HG_i \cap G_{i-1}).$$

But $G_i \triangleleft G_{i-1}$, and $G_i \subseteq HG_i \cap G_{i-1}$, so that by the previous theorem:

$$G_{i-1}/(HG_i \cap G_{i-1}) \cong (G_{i-1}/G_i)/((HG_i \cap G_{i-1})/G_i).$$

The right-hand side is a quotient of a cyclic group, so that either $H_{i-1} = H_i$ or H_{i-1}/H_i is cyclic. Thus G/H is soluble.

Now suppose that $H \triangleleft G$ and that H and G/H are soluble. There exist

$$\{e\} = H_n \triangleleft \cdots \triangleleft H_0 = H, \text{ with } H_{i-1}/H_i \text{ cyclic}$$

and

$$\{H\} = K_m \lhd K_{m-1} \lhd \cdots \lhd K_0 = G/H, \text{ with } K_{j-1}/K_j \text{ cyclic.}$$

Let $q : G \to G/H$ be the quotient mapping, and let $G_j = q^{-1}(K_j)$ for $0 \leqslant j \leqslant m$. If $g_j \in G_j$ and $g_{j-1} \in G_{j-1}$

$$q(g_{j-1}^{-1} g_j g_{j-1}) = (q(g_{j-1}))^{-1} q(g_j) q(g_{j-1}) \in K_j,$$

since $K_j \lhd K_{j-1}$, and so $g_{j-1}^{-1} g_j g_{j-1} \in G_j$. Thus $G_j \lhd G_{j-1}$. By the previous theorem:

$$G_{j-1}/G_j \cong (G_{j-1}/H)/(G_j/H) = K_{j-1}/K_j,$$

which is cyclic. Thus the series

$$\{e\} = H_n \lhd \cdots \lhd H_0 = H = G_m \lhd G_{m-1} \lhd \cdots \lhd G_0 = G$$

shows that G is soluble.   □

What can we say about a transitive soluble subgroup G of Σ_p, where p is a prime number?

Theorem 1.25 *Suppose that G is a soluble transitive subgroup of Σ_S, where $|S|$ is a prime number p. Then G contains a cyclic group H of order p. Label S so that H is the abelian group $\mathbb{Z}_p$. Then G is a subgroup of the affine group $W_{\mathbb{Z}_p}$. Thus $|G| = pk$, where k divides $p - 1$.*

We need a lemma.

Lemma 1.26 *Suppose that G is a subgroup of Σ_p, and that $\mathbb{Z}_p \subseteq G \subseteq W_p$. Then the normalizer $N(G)$ in Σ_p is contained in W_p.*

Proof As before, if $x \in \mathbb{Z}_p$ we define $\alpha(x) = x + 1 \pmod{p}$ and define $\sigma_{a,b}(x) = ax + b \pmod{p}$. Suppose that $\tau \in N(G)$ and $x \in \mathbb{Z}_p$. Then $\tau \alpha \tau^{-1} \in G$, so that $\tau \alpha \tau^{-1} = \sigma_{a,b}$, for some a, b. Now $\tau \alpha \tau^{-1}$ has order p, and so it permutes the points of $\mathbb{Z}_p$ cyclically. Thus the equation $\tau \alpha_1 \tau^{-1}(x) = ax + b = x \pmod{p}$ has no solution. This implies that $a = 1$ and that $b \neq 0$, so that $\tau \alpha \tau^{-1}(x) = x + b$, and $\tau(x + 1) = \tau \alpha(x) = \tau \alpha \tau^{-1} \tau(x) = \tau(x) + b$. Thus $\tau(x) = bx + \tau(0)$, so that $\tau = \sigma_{b,\tau(0)} \in W_p$.   □

Proof of Theorem 1.25 Suppose that H is a proper normal subgroup of G. If $x \in H$, let $O_H(x)$ be the orbit of x. If $y \in S$, there exists $\sigma \in G$ such that $\sigma(x) = y$. Now suppose that $x' \in O_H(x)$, so that there exists $\tau \in H$ such that $\tau(x) = x'$. Then $\sigma(x') = \sigma\tau(x) = \sigma\tau\sigma^{-1}(y) = \tau^{-1}(x') = \tau(y)$, so that $\sigma(O_H(x)) \subseteq O_H(y)$. Similarly, $\sigma^{-1}(O_H(y)) \subseteq O_H(x)$, so that any two H orbits have the same number of elements. This cannot be 1, since H is transitive; thus $|O_x| = p$, so that H is also transitive.

Suppose now that $\{e\} = G_n \lhd G_{n-1} \lhd \cdots \lhd G_0 = G$, where G_i/G_{i+1} is cyclic for $i = 1, \ldots, n-1$, and $G_{n-1} \neq G_n$. Arguing inductively, we see that G_{n-1} is a transitive cyclic subgroup of Σ_S, generated by σ, say. We label S as $\{0, 1, \ldots, p-1\}$ in such a way that $\sigma(j) = j + 1$ for $0 \le j \le p_2$ and $\sigma(p-1) = p_0$, and identify it with $\mathbb{Z}_p$. Repeated use of Lemma 1.26 now shows that $G_{n-j} \subseteq W_p$, for $2 \le j \le n$. $\qquad\qquad\square$

Thus the proper soluble transitive groups of Σ_5 are the cyclic group $\mathbb{Z}_5$ of order 5, the dihedral group D_5 of order 10 and the affine group $W_{\mathbb{Z}_5}$ of order 20.

The transitive subgroups of Σ_5 play a fundamental role in the study of quintic polynomial equations. So far, we have identified five subgroups of Σ_5 which contain $\mathbb{Z}_5$:

Name	Properties	Size	Normalizer	$|conj_{A_5}|$	$|conj_{\Sigma_5}|$
$\mathbb{Z}_5$	abelian	5	D_5	6	12
D_5	soluble	10	W_5	3	6
W_5	soluble	20	W_5	3	6
A_5	simple	60	Σ_5	1	2
Σ_5	not soluble	120	Σ_5	—	1

Are these the only ones? We shall answer this question in the next section.

Burnside used representation theory to show that a group of order $p^a q^b$, where p and q are prime numbers, is soluble.

Exercises

1.23 Give an example of a group of order $p^a q^b r^c$, where p, q and r are prime numbers, which is not soluble.

1.24 Suppose that p is an odd prime number. Show that a group of order $2p$ is either cyclic or isomorphic to D_p. Show that a group of order $4p$ is soluble.

1.25 Suppose that G is a soluble group of order mn, where m and n are coprime, and suppose that H is a subgroup of order k.

(a) Show that G has a subgroup of order m.

(b) Show that any two subgroups of order m are conjugate.

(c) Show that if k divides m then H is contained in a subgroup of order m.

1.6 *p*-Groups and Sylow Theorems

A group is a *p-group* if its order is p^a, where p is a prime number and $a \geq 1$. Throughout this section, p will be a prime number and a a positive integer.

Theorem 1.27 *If G is a p-group, then its centre $Z(G)$ is a p-group; it has more than one element.*

Proof   If C is a conjugacy class in G, then $|C|$ divides $|G|$, so that either p divides $|C|$, or $|C| = 1$. It follows that $|Z(G))| = |\{g \in G : |conj(g)| = 1\}|$ is a multiple of p. Since $Z(G)$ is a subgroup of G, it must be a p-group.   □

Corollary 1.28 *If G is a p-group of order p^a and $1 \leq b \leq a$, then G contains a subgroup of order p^b.*

Proof   We prove this by induction on a. If $a = 1$, the result is trivially true. Assume that the result holds for all $a' < a$. Then $Z(G)$ is an abelian p-group, of order p^c, say. If $c \geq b$, it follows easily from the structure of finite abelian groups that $Z(G)$ contains a subgroup of order p^b. Otherwise, $G/Z(G)$ is a p-group of order p^{a-c}; by the inductive hypothesis, it contans a p-subgroup $H/Z(G)$ of order p^{b-c}. Then H is a p-subgroup of G of order p^b.   □

Theorem 1.29 *A p-group G is nilpotent, and is therefore soluble.*

Proof   Suppose that $|G| = p^a$. We prove the result by induction on a. If $a = 1$, G is cyclic, and so is nilpotent. Suppose that the result holds for $1 \leq c < a$. Now $Z(G) \lhd G$ and $|Z(G)| = p^b$, where $1 \leq b \leq a$ (Theorem 1.27). If $a = b$ there is nothing to prove. Otherwise, $|G/Z(G)| = p^{a-b}$, so that $G/Z(G)$ is nilpotent, by the inductive hypothesis. There is therefore a central series $(G_i/Z(G))_{i=0}^r$ (where $G_r = Z(G)$). If $h \in G_i$ and $g \in G$ then $[h,g]Z(G) = [hZ(G), gZ(G)] \in G_{i+1}Z(G)$, so that $[h,g] \in G_{i+1}$. Thus if we put $G_{r+1} = \{e\}$, we obtain a central series for G.   □

If G is a group of order $p^a r$, where p does not divide r, then a subgroup of order p^a is called a *Sylow p-subgroup*. The theorems that follow are known as *Sylow theorems*. They were proved by the Danish mathematician Ludvig Sylow in 1872. The proofs given here are due to Helmut Wielandt.

Theorem 1.30 *Suppose that G is a group of order $p^a r$, where p does not divide r. Then G contains a Sylow p-subgroup. If v_G is the number of Sylow p-subgroups, then $v_G = 1 \pmod{p}$.*

Proof   Let A be the set of subsets of G of size p^a. We must show that at least one of them is a subgroup of G. Since

$$|A| = \binom{p^a r}{p^a} = \prod_{j=0}^{p^a-1} \frac{p^a - j}{j},$$

p does not divide $|A|$. We show that $|A| = \nu_G$ (mod p).

If $K \in A$ and $g \in G$, let $\pi(g)(K) = Kg$. Then π is a homomorphism of G into Σ_A. If $K \in A$, let $O(K) = \{Kg : g \in G\}$ be the orbit to which K belongs, and let $S(K) = \{g \in G : Kg = K\}$ be the *stabilizer* of K. Then $S(K)$ is a subgroup of G, and $|G| = |O(K)|.|S(K)|$.

Now $K = \cup_{k \in K} kS(K)$ is the union of left cosets of $S(K)$; these are either equal or disjoint, and each have size $|S(K)|$. Thus $p^a = |K| = t|S(K)|$ for some t, and $|S(K)| = p^{b_K}$ for some $0 \le b_K \le a$.

Let $A_1 = \{K \in A : b_K = a\}$ and $A_2 = \{K \in K : b_K < a\}$.

If $K \in A_1$, then $|O(K)| = 1$; $O(K)$ is a singleton and $O(K) = kS(K)$ for some (any) $k \in K$. But then $H = kS(K)k^{-1}$ is a conjugate subgroup of $S(K)$ which is a Sylow p-subgroup. Therefore $K = k^{-1}Hk$ is also a Sylow p-subgroup.

Conversely, if H is a Sylow p-subgroup of G, then $H \in A_1$, so that $|A_1| = \nu_G$. But if $K \in A_2$ then $|O(K)| = p^{a-b_K}$, so that $|A| = \nu_G$ (mod p).

It remains to show that $|A| = 1$ (mod p). Let P be the abelian group $Z_{p^a} \times Z_r$, and let A_P be the collection of subsets of P of size p^a, so that $|A| = |A_P|$. But $\nu_P = 1$, so that $|A| = \nu_p = 1$ (mod p). □

Corollary 1.31 *Suppose that G is a soluble transitive subgroup of Σ_S, where S is a finite set with a prime number p of elements. Then G contains a cyclic group H of order p.*

Proof Since G is transitive and soluble, G contains a cyclic subgroup of order p, and so G contains a Sylow p-subgroup H. Since p^2 does not divide $p! = |\Sigma_p|$, $|H| = p$, and H is cyclic. □

In this case, we can number S in such a way that $\alpha \in H$, where $\alpha(j) = j+1$ (mod p).

Theorem 1.32 *(i) Suppose that G is a group of order $p^a r$, where p does not divide r, and that H is a p-subgroup of G. Then H is contained in a Sylow p-subgroup.*

(ii) Any two Sylow p-subgroups are conjugate.

Proof (i) Let S be a Sylow p-subgroup of G. We let H act on the set B of cosets of S: if $h \in H$ and $Sg \in B$, let $\theta(h)(Sg) = Sgh$. Since $|B| = r$ and every orbit of π has size p^c, for some $c \ge 0$, there exists an orbit of size 1; that is, there exists g_0 such that $Sg_0h = Sg_0$ for all $h \in H$. Thus $H \subseteq g_0^{-1}Sg_0$, a Sylow p-subgroup conjugate to S.

(ii) If H is a Sylow p-subgroup, then, arguing as above, $H = g_0^{-1}Sg_0$, so that H and S are conjugate. $\square$

Corollary 1.33 *A Sylow p-subgroup of G is normal if and only if it is the unique Sylow p-subgroup of G.*

We can now answer the questions left open in the previous section.

Theorem 1.34 *If G is a subgroup of A_5 which contains $\mathbb{Z}_5$, then $G = \mathbb{Z}_5, D_5$ or A_5.*

Proof   Simple calculations show that $N_{A_5}(\mathbb{Z}_5)$, the normalizer of $\mathbb{Z}_5$ in A_5, is equal to D_5, and that A_5 is generated by the 5-cycles. Consider the Sylow subgroups of G; they have five elements. If there is only one, then $\mathbb{Z}_5 \triangleleft G \subseteq N_{A_5}(\mathbb{Z}_5)$, so that $G = \mathbb{Z}_5$ or D_5. Otherwise, all six such subgroups are in G, so that $G = A_5$. $\square$

Theorem 1.35 *If G is a subgroup of Σ_5 which contains $\mathbb{Z}_5$ and is not contained in A_5, then $G = W_5$ or Σ_5.*

Proof   If $G \neq \Sigma_5$, let $H = G \cap A_5$. Then $H = \mathbb{Z}_5$ or D_5, $|G| = 2|H|$, and so $H \triangleleft G$. Since $N_{\Sigma_5}(\mathbb{Z}_5) = D_5$ and $N_{\Sigma_5}(D_5) = W_5$, the result follows. $\square$

Exercises

1.26   Verify the easy calculations of the two previous theorems.

1.27   Show that if G is a p-group of order p^2 then G is abelian. Does the result hold for groups of order p^3?

1.28   How many Sylow p-subgroups does Σ_p possess?
       Deduce Wilson's theorem that $(p-1)! = -1 \pmod{p}$.

1.29   Show that there are three Sylow 2-groups in Σ_4, each isomorphic to the dihedral group D_4.

1.30   What are the Sylow 3-groups in Σ_4?

1.31   Classify the groups of order 12.

2

Integral Domains

The second main topic of Galois theory is the study of polynomials. The collection of all polynomials with integral coefficients forms an *integral domain*, and integral domains provide an appropriate setting for the study of divisibility and factorization.

2.1 Commutative Rings with 1

A *commutative ring with a* 1 is a non-empty set R with two laws of composition, addition and multiplication. Under addition, R is an abelian group, written additively, with neutral element 0. Multiplication is a binary operation, written $(r,s) \to rs$, which satisfies:

(a) $rs = sr$, $(rs)t = r(st)$ and $r(s + t) = rs + rt$ for all r,s,t in R;
(b) there exists an identity element $1 \neq 0$ such that $1r = r$ for all r in R.

If $rs = t$, r and s *divide* t, and we write $r|t$ and $s|t$. If $r|1$, r is called a *unit*. The set of units is a commutative group under multiplication, with 1 as identity element. If $rs = 1$, then s is called the inverse of r, and is written as r^{-1}. A *unit* in a ring is an element with an inverse. The set of units is a group, under multiplication.

The integers $\mathbb{Z}$, with their usual operations of addition and multiplication, form a commutative ring with a 1. The units in $\mathbb{Z}$ are 1 and -1. If R is a commutative ring with a 1 and S is a set, the set $R(S)$ of all R-valued functions on S, with pointwise addition and multiplication, forms a commutative ring with a 1.

Algebraists study non-commutative rings (for example, rings of matrices) and rings without a 1. Such rings will not concern us.

25

From now on, we shall abbreviate 'commutative ring with a 1' to 'ring'.

A subset S of a ring R is a *subring* if it is a ring with the operations of addition and multiplication inherited from R. The identity element of S need not be the identity element of R.

An element r of a ring is *nilpotent* if $r^n = 0$ for some $n \in \mathbb{N}$.

Exercises

2.1   Let $\omega = (-1 + i\sqrt{3})/2 \in \mathbb{Q}$. Show that $R = \{m + \{\omega\}n : m, n \in \mathbb{Z}\}$ is a subring of $\mathbb{C}$. What are the units in R?

2.2   Show that a nilpotent element r is not a unit, but that $1 + sr$ is a unit, for each $s \in R$.

2.3   Let S be a set of prime numbers. Let $\mathbb{Q}_S = \{(n/t) : n \in \mathbb{Z}, t$ has no prime factor in S$\}$. Show that $\mathbb{Q}_S$ is a subring of $\mathbb{Q}$, and that every subring of $\mathbb{Q}$ is of this form. When is $\mathbb{Q}_S$ a maximal proper subring of $\mathbb{Q}$?

2.2 Polynomials

Suppose that R is a ring. Let

$$R[x] = \{(a_0, a_1, \ldots, a_n, 0, 0, \ldots) : n \in \mathbb{N}, a_i \in \mathbb{R}\}$$

be the space of all sequences in R with only finitely many non-zero terms, with addition defined coordinate by coordinate, and multiplication as follows: if $a = (a_i)_{i=0}^{\infty}$ and $b = (b_j)_{j=0}^{\infty}$ then $ab = c = (c_k)_{k=0}^{\infty}$, where $c_k = \sum\{a_i b_j : i + j = k\}$.

Exercise

2.4   Verify that this satisfies the conditions for being a ring (with 1 equal to $(1, 0, 0, \ldots)$).

Now let $x = (0, 1, 0, 0, \ldots)$: x is called an *indeterminate* or *variable*. Then $x^r = (0, 0, \ldots, 0, 1, 0, \ldots)$, so that if $a = (a_0, a_1, \ldots, a_n, 0, 0, \ldots)$ then $a = a_0 + a_1 x + \cdots + a_n x^n$; addition and multiplication correspond to the usual multiplication of polynomials, and so we can identify $R[x]$ with the *ring of polynomials* with coefficients in R. If $a \in R[x]$, the smallest n for which $a_m = 0$ for $m > n$ is the *degree* of a, and a is *monic* if $a_n = 1$. If r has degree 0, r is a *constant*.

Arguing inductively, we can consider polynomials in several variables: we define $R[x_1, \ldots, x_n] = R[x_1, \ldots, x_{n-1}][x_n]$. The elements $x_1, \ldots, x_n$ are called *indeterminates*.

In the same way, we can construct the ring consisting of all formal power series in one or more variables, but we shall not need to consider these in what follows.

2.3 Homomorphisms and Ideals

Suppose that R and S are two rings. A mapping ϕ from R to S is called a *ring homomorphism* if

(a) $\phi(r_1 + r_2) = \phi(r_1) + \phi(r_2)$,
(b) $\phi(r_1 r_2) = \phi(r_1)\phi(r_2)$ and
(c) $\phi(1_R) = 1_s$

for r_1, r_2 in R. Thus a ring homomorphism is a group homomorphism for the additive groups $(R, +)$ and $(S, +)$ which respects multiplication.

A homomorphism which is one-to-one is called a *monomorphism*, one which is onto is called an *epimorphism* and one which is both is called an *isomorphism*. An isomorphism of a ring R onto itself is called an *automorphism*.

The image $\phi(R)$ is a subring of S. The kernel $\phi^{-1}(\{0\})$ is not (since $1_R \notin \phi^{-1}(\{0\})$), but has rather different properties. A non-empty subset J of a ring R is said to be an *ideal* if the following conditions hold:

(i) if r and s are in J, so is $r + s$;
(ii) if $r \in R$ and $s \in J$, then $rs \in J$.

Note that if $s \in J$, then $-s = (-1)s \in J$, so that J is a subgroup of the additive group $(R, +)$. The ring R is an ideal in R; all ideals other than R are called *proper ideals*.

Let us consider some examples.

1. The sets $n\mathbb{Z}$ are ideals in the ring $\mathbb{Z}$, and any ideal of $\mathbb{Z}$ is of this form.

2. Suppose that A is a non-empty subset of a ring R. We denote by $<A>$ the intersection of all ideals which contain A. $<A>$ is called the *ideal generated by A*. Further:

$$<A> = \{r \in R: r = r_1 a_1 + \cdots + r_n a_n; r_i \in R, a_i \in A\},$$

for every element in the set on the right-hand side must be in $<A>$, and it is easy to see that this set is an ideal which contains A.

3. We write $(a_1, \ldots, a_n)$ for $(\{a_1, \ldots, a_n\})$. An ideal (a) generated by a single element is called a *principal* ideal. (a) consists of all multiples of a by elements of R.

If ϕ is a ring homomorphism from R into S, and $\phi(r) = \phi(s) = 0$, then

$$\phi(r + s) = \phi(r) + \phi(s) = 0.$$

Also if $t \in R$, $\phi(tr) = \phi(t)\phi(r) = \phi(t)0 = 0$; thus the kernel is an ideal. As $\phi(1_R) \neq 0$, the kernel is a proper ideal.

If J is a proper ideal in R, J is a normal subgroup of $(R, +)$; we can construct the quotient group R/J. We can also define the product of two (right) cosets: if C_1 and C_2 are two cosets, we define

$$C_1 C_2 = \{c_1 c_2 + j : c_1 \in C_1, c_2 \in C_2, j \in J\}.$$

If $c_1' c_2' + j'$ and $c_1 c_2 + j$ are elements of $C_1 C_2$:

$$(c_1' c_2' + j') - (c_1 c_2 + j) = c_1'(c_2' - c_2) + c_2(c_1' - c_1) + (j' - j) \in J,$$

so that $C_1 C_2$ is a coset. It is straightforward to verify that with these operations R/J is a ring, with unit $J + 1_R$, and that the quotient map $q : R \to R/J$ is a ring homomorphism, with kernel J. As an example, the quotient $\mathbb{Z}_n = \mathbb{Z}/n\mathbb{Z}$ is a ring, the *ring of integers* (mod n). Just as for groups, we have an isomorphism theorem:

Theorem 2.1 (Quotient ring) *Suppose that ϕ is a ring homomorphism from a ring R to a ring S, with kernel J. There is a ring isomorphism $\tilde{\phi}$ from R/J onto $\phi(R)$ such that $\phi = \tilde{\phi} \circ q$.*

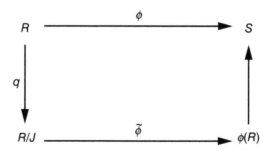

Proof ϕ is a group homomorphism from $(R, +)$ to $(S, +)$, so that by the first isomorphism theorem for groups there is a group isomorphism $\tilde{\phi} : R/J \to \phi(R)$ such that $\phi = \tilde{\phi}q$. If C_1 and C_2 are two cosets of J, we can write

$C_1 = q(x_1) = J + x_1$, $C_2 = q(x_2) = J + x_2$ for some x_1 and x_2 in R. Then $C_1 C_2 = q(x_1 x_2) = J + x_1 x_2$, and so

$$\tilde{\phi}(C_1 C_2) = \tilde{\phi}(q(x_1 x_2)) = \phi(x_1 x_2) = \phi(x_1)\phi(x_2)$$
$$= \tilde{\phi}(q(x_1))\tilde{\phi}(q(x_2)) = \tilde{\phi}(C_1)\tilde{\phi}(C_2).$$

Similarly

$$\tilde{\phi}(J + 1_R) = \tilde{\phi}q(1_R) = \phi(1_R) = 1_S. \qquad \square$$

Theorem 2.2 *Suppose that R is a ring with unit element 1_R and zero element 0_R. Then there is a unique ring homomorphism ϕ from $\mathbb{Z}$ into R. Either ϕ is injective or $\phi^{-1}(0_R) = < n >$, for some $n > 0$.*

Proof For if we define $\phi(0) = 0_R$ and $\phi(1) = 1_R$, define $\phi(n + 1)$ inductively as $\phi(n) + 1_R$ for $n > 0$, and define $\phi(-n) = -\phi(n)$, it is straightforward to verify that $\phi(mn) = \phi(m)\phi(n)$ for all m and n, so that ϕ is a ring homomorphism, and it is equally straightforward to verify that any homomorphism is of this form. $\qquad \square$

The image $\phi(\mathbb{Z})$ is (misleadingly) called the *prime ring* of R. If ϕ is injective, it is isomorphic to $\mathbb{Z}$, and we say that R has *characteristic 0*. If not, its kernel $\phi^{-1}(0_R)$ is a non-zero ideal $< n >$ in $\mathbb{Z}$ for some $n > 1$, and the image $\phi(\mathbb{Z})$ is isomorphic to $\mathbb{Z}_n$, by Theorem 2.1. In this case, we say that R has *characteristic n*.

Exercises

2.5 Give an example of a ring R for which $1 + x = f(x)g(x)$ in $R[x]$, where $f(x)$ and $g(x)$ are not units.

2.6 Let $(I_n)_{n=1}^{\infty}$ be a strictly increasing sequence of proper ideals in a ring R. Show that $\cup_{n=1}^{\infty} I_n$ is a proper ideal in R.

2.7 Suppose that I and J are ideals in a ring R. Show that $I \cap J$ is also an ideal. Construct a monomorphism from $R/(I \cap J)$ into $R/I \times R/J$. When is it an isomorphism?

2.8 Construct for each positive integer n an ideal in $\mathbb{Z}[x]$ which is generated by n elements and is not generated by fewer than n elements.

2.4 Integral Domains

An *integral domain* is a commutative ring R with a 1 which has the property that if $r, s \in R$ and if $r \neq 0$ and $s \neq 0$ then $rs \neq 0$.

Let us give some examples of integral domains.

1. The integers $\mathbb{Z}$, with the usual operations of addition and multiplication. This is the fundamental example, and other integral domains are generalizations of it.

2. The *Gaussian integers* $\mathbb{Z} + i\mathbb{Z} = \{m + in : m, n \in \mathbb{Z}, i^2 = -1\}$.

3. $\mathbb{Z} + \sqrt{k}\mathbb{Z} = \{m + \sqrt{k}n : m, n, k \in \mathbb{Z}, k \text{ not a square}\}$.

4. If R is an integral domain then $R[x]$ is an integral domain. For if a is a non-zero polynomial of degree m and b is a non-zero polynomial of degree n and $c = ab$ and $l = m + n$ then $c_l = a_m b_n \neq 0$. Thus if R is an integral domain, so is $R[x_1, \ldots, x_n]$.

Exercise

2.9   Let $J = j \in \mathbb{Q} : j = r/s$, with s odd. Show that J is an integral domain. What are the invertible elements of J? What are the ideals in J? If I is an ideal in J, describe the quotient J/I.

Polynomials over integral domains behave much as polynomials with coefficients in $\mathbb{Z}$. Polynomials in several variables are more complicated.

Proposition 2.3 *Suppose that R is an integral domain, that $f \in R[x]$ is a monic polynomial of degree n and $g \in R[x]$. Then there exist $q, r \in R[x]$ with the degree of r less than n such that $g = qf + r$.*

Proof   Familiar long division.   □

Suppose that R is an integral domain and that $(r_1, \ldots, r_k) \in R^k$. If $f \in R[x_1, \ldots, x_k]$, let $e(r_1, \ldots, r_k)(f) = f(r_1, \ldots, r_k)$. Then e is a ring homomorphism of R^n into R, the *evaluation map*. If $e(r_1, \ldots, r_k)(f) = 0$ we say that $(r_1, \ldots, r_k)$ is a *zero* of f. When $k = 1$ it is also called a *root* of f.

Corollary 2.4 *If r is a root of $f \in R[x]$, where R is an integral domain, then $(x - r) | f$.*

If r is a root of $f \in R[x]$, the *multiplicity* of r is the largest integer k such that $(x - r)^k | f$.

Corollary 2.5 *Suppose that $f \in R[x]$ has degree n and if $\{r_1, \ldots, r_j\}$ are the roots of f, with multiplicities $\{k_1, \ldots, k_j\}$, then $\sum_{i=1}^{j} k_i \leq n$.*

Exercise

2.10   Suppose that R is an integral domain with identity element 1_R, and that S is a subring of R. Show that $1_R \in S$.

The polynomial $x_1^2 + x_2^2 - x_3^2$ has infinitely many zeros in $\mathbb{Z}[x_1, x_2, x_3]$. We can however say something about the zeros of a polynomial in $R[x_1, \ldots, x_k]$ when R has infinitely many elements.

Proposition 2.6 *Suppose that R is an integral domain with infinitely many elements. If $f \in R[x_1, \ldots, x_k]$ and $f \neq 0$ then there exists $(r_1, \ldots, r_k) \in R^k$ which is not a zero of f.*

Proof We prove this by induction on k. The result is true if $k = 1$, by Corollary 2.5. Suppose that it is true for $k - 1$ and that f is a non-zero element of $R[x_1, \ldots, x_k]$. We can consider f as an element of $R[x_1][x_2, \ldots, x_k]$. By the inductive hypothesis, there exists $(r_2, \ldots, r_k)$ such that $f(x, r_2, \ldots, x_k) \neq 0$, and there exists r_1 such that $f(r_1, \ldots, r_k) \neq 0$. $\square$

The quotient of an integral domain by an ideal need not be an integral domain.

Proposition 2.7 *The quotient $\mathbb{Z}_n = \mathbb{Z}/ < n >$ is an integral domain if and only if n is a prime number p.*

Proof Let $q : \mathbb{Z} \to \mathbb{Z}_n$ be the quotient map, and let $\bar{a} = q(a)$. If $n = ab$ is not a prime number, $\bar{a} \neq 0$ and $\bar{b} \neq \bar{0}$, but $\bar{a}\bar{b} = \bar{0}$. If n is a prime number p and $\bar{a} \neq 0$ and $\bar{b} \neq \bar{0}$, then $p \nmid a$ and $p \nmid b$, so that $p \nmid ab$. Thus $\bar{a}\bar{b} \neq 0$. $\square$

Corollary 2.8 *An integral domain either has characteristic 0 or a prime number p.*

Exercises

2.11 Suppose that R is an integral domain with characteristic k. Show that, when R is considered as an additive group, every non-zero element has order k (if $k > 0$) or infinite order (if $k = 0$).

2.12 Suppose that R is an infinite ring such that R/I is finite for each non-trivial ideal I. Show that R is an integral domain.

2.5 Fields and Fractions

Suppose that R is a ring. An element a of R is *invertible*, or a *unit*, if it has a multiplicative inverse; that is, there is an element a^{-1} such that $aa^{-1} = 1$. If so, the inverse is unique, for if a' is an inverse then $a' = a'(aa^{-1}) = (a'a)a^{-1} = a^{-1}$. If a_1 and a_2 are units, so is a_1a_2 (with inverse $a_2^{-1}a_1^{-1}$), so that the set of units in R forms a group U_R under multiplication.

Exercise

2.13 If $k \in \mathbb{Z}_n$, then $k \in U_n$, the group of units of $\mathbb{Z}_n$ if and only if the numbers k and n have no common factor.

A ring F is a *field* if every non-zero element of F is a unit. The study of fields is an essential part of Galois theory.

Examples of fields are the sets $\mathbb{Q}$ of rational numbers, $\mathbb{R}$ of real numbers and $\mathbb{C}$ of complex numbers.

Proposition 2.9 *A ring R is a field if and only if $\{0\}$ and R are the only ideals in R.*

Proof Suppose that R is a field, that J is an ideal in R other than $\{0\}$ and that a is a non-zero element of J. If $b \in R$ then $b = a(a^{-1}b) \in J$, and so $J = R$.

Conversely, suppose that $\{0\}$ and R are the only ideals in R. If $a \neq 0$ then $<a> = R$, so that there exists b such that $ab = 1$, and a is invertible. □

Corollary 2.10 *If ϕ is a ring homomorphism from a field F into a ring R then ϕ is a monomorphism.*

Proof For the kernel $\phi^{-1}(\{0\})$ is an ideal in F. □

Corollary 2.11 *If J is an ideal in a ring R, then the quotient R/J is a field if and only if J is a maximal proper ideal in R. $\mathbb{Z}_n$ is a field if and only if n is a prime number.*

Proof For if $q : R \to R/J$ is the quotient map then I is an ideal in R/J if and only if $q^{-1}(I)$ is an ideal in R. □

Suppose that F is a field. A *subfield* of F is a subset of F which is a field under the operations inherited from F. Any subfield must therefore contain 0 and 1. The intersection of subfields is a subfield F_0. The intersection of all subfields is therefore a subfield, the *prime* subfield of F. F is a commutative ring with a 1, and we can therefore consider the ring homomorphism $\phi : \mathbb{Z} \to F$ described earlier in the chapter. Clearly $\phi(\mathbb{Z}) \subseteq F_0$. If ϕ has non-zero characteristic n, then, since F_0 is an integral domain, n must be a prime number, and F_0 is isomorphic as a field to $\mathbb{Z}_p$.

On the contrary, if ϕ has characteristic 0 then $\phi(\mathbb{Z})$ is isomorphic to $\mathbb{Z}$. If $r/s \in \mathbb{Q}$, let $\phi(q) = \phi(r)\phi(s)^{-1}$. This is well defined, since if $q = r'/s'$ then $rs' = r's$ so that $\phi(r)\phi(s') = \phi(r')\phi(s)$ and $\phi(r)\phi(s)^{-1} = \phi(r')\phi(s')^{-1}$. Thus ϕ is properly defined, and it is easy to show that $\phi(\mathbb{Q})$ is a subfield of F and that ϕ is a field isomorphism of $\mathbb{Q}$ onto S, $\phi(\mathbb{Q})$. Clearly every element of $\phi(\mathbb{Q})$ is in F_0, so that $\phi(\mathbb{Q}) = F_0$. Summing up:

Theorem 2.12 *Suppose that F is a field with prime subfield F_0. If F has characteristic 0, then F_0 is isomorphic as a field to $\mathbb{Q}$. Otherwise, F has characteristic p, where p is a prime number, and F_0 is isomorphic to the finite field $\mathbb{Z}_p$.*

Many integral domains and fields that we shall consider are subsets of $\mathbb{C}$, with the same operations of addition and multiplication. We shall call these *numerical integral domains* and *number fields*.

We now show how to construct a field from an integral domain, in the same way that the field $\mathbb{Q}$ of rational fractions is constructed from $\mathbb{Z}$.

Suppose that R is an integral domain; let $R^* = R \setminus \{0\}$. Intuitively, a fraction is an expression of the form a/b, with $a \in R$ and $b \in R^*$, but a fraction can be represented by many such expressions. We therefore proceed as follows. We define an equivalence relation $\sim$ on $R \times R^*$ by setting $(a_1, b_1) \sim (a_2, b_2)$ if $a_i b_2 = a_2 b_1$. It is immediate that $(a, b) \sim (a, b)$ and that if $(a_1, b_1) \sim (a_2, b_2)$ then $(a_2, b_2) \sim (a_1, b_1)$. Finally, if $(a_1, b_1) \sim (a_2, b_2)$ and $(a_2, b_2) \sim (a_3, b_\jmath)$, then $a_1 b_2 b_3 = a_2 b_1 b_3 = a_3 b_1 b_2$, so that $(a_1 b_3 - a_3 b_1) b_2 = 0$; thus, since R is an integral domain, $a_1 b_3 - a_3 b_1 = 0$ and $(a_1, b_1) \sim (a_3, b_3)$. Consequently $\sim$ is an equivalence relation. Let F be the collection of equivalence classes; we denote the class to which (a, b) belongs by (a/b). We define operations of addition and multiplication by

$$a_1/b_1 + a_2/b_2 = (a_1 b_2 + a_2 b_1)/(b_1, b_2) \text{ and } (a_1/b_1)(a_2, b_2) = (a_1 a_2)/(b_1 b_2).$$

Exercises

2.14 Check that these do not depend upon the choice of representatives, that under these operations F is a field, with identity $1/1$, zero element $0/1$ and that the multiplicative inverse of a non-zero a/b equals b/a.

2.15 Show that the mapping $a \to a/1$ is a ring isomorphism of R into F: $a/1 + b/1 = (a + b)/1$ and $(a/1).(b/1) = (ab)/1$ for all $a, b \in R$.

    F is called the *field of fractions* of R. If R is an integral domain with field of fractions F, the mapping $i : R \to F$ defined by $i(r) = r/1$ is a monomorphism of R into F.

2.16 Suppose that R is an integral domain with field of fractions F, and that j is an automorphism of R. Show that there is a unique automorphism j' of F such that $j'i = ij$ (where $i(r) = r/!$). What is it?

2.17 Identify the field of fractions of the ring $\mathbb{Z} + i\mathbb{Z}$ of Gaussian integers with a subfield of $\mathbb{C}$.

To end this section, let us consider some groups that arise in the study of fields. Suppose that F is a field, and that F^* is the set of non-zero elements of F. If $a \in F^*$ let $\mu_a(f) = af$, for $f \in F$, and if $b \in F$ let $\alpha_b(f) = f + b$. Then $M_F = \{\mu_a : a \in F^*\}$ is an abelian group of permutations of F, with two orbits, the fixed point $\{0\}$ and F^*, and $A_F = \{\alpha_b : b \in F\}$ is a transitive abelian group of permutations of F. The *affine group* W_F is then the group generated by M_F and A_F; since $\alpha_b \mu_a(f) = af + b = \mu_a \alpha_{b/a}$, every element of W_F can be written as $\gamma_{(a,b)}$, where $\gamma_{(a,b)}(f) = af + b$. The map $w_{(a,b)} \to \mu_a$ is a monomorphism of W_F onto M_F, with kernel A_F, so that $A_F \lhd W_F$ and $W_F/A_F \sim M_F$. Since A_F and M_F are both abelian, it follows that W_F is nilpotent, and therefore is soluble.

Theorem 2.13 *Suppose that F is a field, with multiplicative group (F^*, M_F) of non-zero elements. If G is a finite subgroup of F^*, then G is cyclic.*

Proof   Let λ be the exponent of G. Then $x^\lambda = 1$ for all $x \in G$. But $x^\lambda - 1$ has at most λ roots in F^*, and so $|G| \leq \lambda$. But λ divides $|G|$, so that $\lambda = |G|$, and G is cyclic.   □

Corollary 2.14 *Suppose that F is a finite field of order q. Then (F^*, M_F) is cyclic.*

Theorem 2.15 *Suppose that G is a soluble transitive subgroup of S, where $|S|$ is a prime number p. Then G contains a cyclic group H of order p. Label S so that H is the abelian group $\mathbb{Z}_p$. Then G is a subgroup of the affine group $W_{\mathbb{Z}_p}$. Thus $|G| = pk$, where k divides $p - 1$.*

Proof   Suppose that H is a proper normal subgroup of G. If $x \in H$, let $O_H(x)$ be the orbit of x. If $y \in X$, there exists $\sigma \in G$ such that $\sigma(x) = y$. Now suppose that $x' \in O_x$, so that there exists $\tau \in H$ such that $\tau(x) = x'$. Then $\sigma(x') = \sigma\tau(x) = \sigma\tau\sigma^{-1}(y)\tau^{-1}(x') = \tau(y)$, so that $\sigma(0_x) \subseteq O_y$. Similarly $\sigma^{-1}(O_y) \subseteq O_x$, so that any two H orbits have the same number of elements. This cannot be 0, since H is transitive; $|O_x| = p$, so that H is also transitive. Suppose now that $\{e\} = G_n \lhd G_{n-1} \lhd \cdots \lhd G_0 = G$, where G_i/G_{i+1} is cyclic for $i = 1, \ldots, n - 1$, and $G_{n-1} \neq G_n$. Arguing inductively, we see that G_{n-1} is a transitive cyclic subgroup of Σ_p, generated by σ, say. We label S as $\{0, 1, \ldots, p - 1\}$ in such a way that $\sigma(j) = j + 1$ for $0 \leq j \leq p_2$ and $\sigma(p - 1) = p_0$, and identify it with $\mathbb{Z}_p$.   □

For example, the proper soluble transitive groups of Σ_5 are the cyclic group $\mathbb{Z}_5$ of order 5, the dihedral group D_5 of order 10 and the affine group $W_{\mathbb{Z}_5}$ of order 20.

2.6 The Ordered Set of Ideals in an Integral Domain

Suppose that R is an integral domain. We now consider the set P of proper ideals of R as an ordered set, ordered by inclusion. We begin with a general result

A proper ideal I in a ring is a *maximal* proper ideal if it is not contained in a larger proper ideal: if J is a proper ideal which contains I, then $J = I$.

Theorem 2.16 *Suppose that J is a proper ideal of a ring R. Then there exists a maximal proper ideal which contains J.*

Proof We use Zorn's lemma. Let P_J denote the collection of proper ideals of R which contain J. We order P_J by inclusion. If C is a chain in P_J, let

$$I_0 = \bigcup \{I : I \in C\}.$$

If a_1 and $a_2 \in I_0$, there exist I_1 and I_2 in C such that $a_1 \in I_1$ and $a_2 \in I_2$. As C is a chain, either $I_1 \subseteq I_2$ or $I_2 \subseteq I_1$. If $I_1 \subseteq I_2$, then $a_1 \in I_2$ and so $a_1 + a_2 \in I_2$. As $I_2 \subseteq I_0, a_1 + a_2 \in I_0$. A similar argument applies if $I_2 \subseteq I_1$. More trivially, if $a \in I_0$ and $r \in R$, then $a \in I_1$ for some I_1 in C. Then $ra \in I_1$, and so $ra \in I_0$. Thus I_0 is an ideal.

Further, if $I \in C$, $1 \notin I$, since I is proper. Thus $1 \notin I_0$. Consequently $I_0 \in P_J$; I_0 is an upper bound for C, and so P_J contains a maximal element, by Zorn's lemma. $\square$

In contrast, if R is countable, we can use Theorem A.1, and avoid use of the axiom of choice.

Exercises

2.18 Show that an element of a ring R is invertible if and only if it is contained in no maximal proper ideal in R.

2.19 Suppose that J is a prime ideal in an integral domain R. Show that $J[x]$ is prime in $R[x]$. Show that $J[x]$ is not a maximal ideal in $R[x]$.

In contrast, a proper ideal in an integral domain R need not contain a minimal proper ideal.

Exercise

2.20 Let $R = \mathbb{Z}(x_1, x_2, \ldots)$, where $(x_n)_{n=1}^{\infty}$ is an infinite sequence of variables. Show that there exists an infinite decreasing sequence of proper ideals with no lower bound in P.

There is another condition, concerning all ideals, which is very useful. An integral domain R is *Noetherian* if every ideal is finitely generated: if I is an ideal, then there is a finite set F such that $I = <F>$. We order the proper ideals by inclusion.

Theorem 2.17 *The following are equivalent:*

  (i) *R is Noetherian.*
 (ii) *If $(J_n)_{n=1}^\infty$ is an increasing sequence of ideals in R, there exists n such that $J_m = J_n$ for $m > n$.*
(iii) *If S is a non-empty set of proper ideals, there exists J_0 in S which is maximal in S; if $J \in S$ then J does not strictly contain J_0.*

Proof   Suppose that R is Noetherian and that (J_n) is an increasing sequence of proper ideals. Then $J = \cup_{j=1}^\infty J_i$ is a proper ideal, since $1_R \notin J$. It is therefore generated by a finite subset, and therefore by a subset of some J_n. Then $J_m = J_n$ for $m > n$, so that (ii) holds.

Suppose that S is a non-empty set of proper ideals which does not satisfy (iii). Let $J_0 \in S$. Then there exists $J_1 \in S$ which strictly contains J_0. Repeating the procedure, we obtain a strictly increasing sequence of proper ideals in S, so that (ii) does not hold. Thus (ii) implies (iii).

Suppose that (iii) holds, and that J is a proper ideal in R. Let S be the set of all finitely generated ideals contained in J. Then there exists an ideal J_0 which is maximal in S. If $a \in J$, then $<a> \subset J_0$, so that $J_0 = J$. Thus J is finitely generated, and (iii) implies (i).   □

Corollary 2.18 *(i) A proper ideal in a Noetherian integral domain is contained in a proper maximal ideal.*

Exercise

2.21   Show that an integral domain R is Noetherian if and only if every ideal in R is finitely generated: if I is an ideal in R there exists a finite set A such that $I = <A>$. If R is Noetherian is $R[x]$ Noetherian?

2.7 Factorization

In this and the next section, we suppose that R is an integral domain.

If a is a non-zero element of R which is not a unit, we say that a *factorizes* if $a = bc$, where neither b nor c is a unit. If a does not factorize, we say that a

is *irreducible*; if a is irreducible and $a = bc$, then either b or c is a unit. Thus a *prime number* is an integer greater than 1 which is irreducible in $\mathbb{Z}$.

We characterize irreducibility in terms of ideals. A proper principal ideal is an ideal $< a >= aR$, where a is a non-zero element of R which is not a unit. We say that a and a' are *associates* if $< a >=< a' >$.

Proposition 2.19 *Elements a and a' of an integral domain R are associates if and only if there exists a unit b such that $a' = ab$.*

Proof If a and a' are associates, then $a' = ab$ for some b and $a = a'b'$ for some b', so that $a' = a'b'b$, and $0_R = a'(1_R - b'b)$. Since R is an integral domain, $b'b = 1_R$, and b and b' are units. Conversely, if $a' = ab$, where b is a unit, then $a' \in < a >$, so that $< a' > \subseteq < a >$; but $a = b^{-1}a'$, so that $< a > \subseteq < a' >$, too.   □

Let PP denote the collection of proper principal ideals of R. We order PP by inclusion: $< a > \leq < b >$ if and only if $< a > \subseteq < b >$. It follows from Proposition 2.19 that $< a > \leq < b >$ if and only if there is an element c such that $a = cb$; thus b divides a, which we write as $b|a$.

An element $< a >$ of PP is a *maximal* element of PP if it is not properly contained in another element of PP. A maximal element does not need to be the largest element of PP.

Theorem 2.20 *A non-zero element a of R is irreducible if and only if $< a >$ is a maximal element of PP.*

Proof If a is irreducible, a does not divide 1_R, so that $< a > \in PP$. If $< a > \leq < b >$, then $b|a$, so that $a = bc$; since a is irreducible, c is a unit, and $< b > = < a >$. Thus $< a >$ is maximal.

Conversely, if a is not irreducible and $a = bc$, where b and c are not units, then $< a > \leq < b >$, but $< b > \not\leq < a >$, so that $< a >$ is not maximal.   □

Corollary 2.21 *If every non-zero element of R which is not a unit can be expressed as the product of a finite number of irreducible elements, then every element of PP is contained in a maximal element of PP.*

We now introduce a condition which ensures that every non-zero element of R which is not a unit can be expressed as the product of a finite number of irreducible elements. R is said to satisfy the ascending chain condition for principal ideals (ACCPI) if whenever $I_1 \subseteq I_2 \subset I_2 \subseteq \cdots$ is an increasing sequence of principal ideals then there exists n such that $I_m = I_n$ for all $m \geq n$.

Exercise

2.22 A Noetherian ring satisfies the ACCPI. Show that $\mathbb{Z}[X]$ satisfies the ACCPI, but is not Noetherian.

Theorem 2.22 *If R satisfies the ACCPI then every non-zero element of R which is not a unit can be expressed as the product of a finite number of irreducible elements.*

Proof Suppose that $a = a_1$ is a non-zero element of R which is not a unit. If $< a_1 >$ is maximal, a is irreducible, and there is nothing to prove. Otherwise, there exists $< a_2 >$ such that $< a_1 > \subset < a_2 >$ and $< a_1 > \neq < a_2 >$. If $< a_2 >$ is maximal, we stop. Otherwise, we repeat the procedure, and continue. The process must stop after a finite number of steps, by the ACCPI. Thus there exists $< a_n >$ with $< a_n >$ maximal, and $< a_n > \geq < a >$. Thus a_n is irreducible, and $a_n | a$. Hence, writing $b_0 = a_n$, we can write $a = b_0.b_1$. If b_0 is irreducible, we are done. Otherwise, we repeat, and continue. The process must terminate after a finite number of steps, for otherwise we get a strictly increasing sequence of principal ideals $(b_k)_{k=1}^{\infty}$, contradicting the ACCPI. Thus $a = b_0.b_1 \ldots b_k$ for some k, where the b_i are irreducible. $\qquad\qquad \square$

Exercises

2.23 Show that the polynomial $x^4 + a^2 x^2 + 6x + 30$ is irreducible in $\mathbb{Z}[x]$ for all $\mathbf{a} \in \mathbb{Z}$.

2.24 Find an element of $\mathbb{Z}[i\sqrt{17}]$ which is the product of two irreducible factors and also the product of three irreducible factors.

2.8 Unique Factorization

The numerical integral domain $\mathbb{Z}[i\sqrt{5}] = \{a + i\sqrt{5}b : a, b \in \mathbb{Z}\}$ is Noetherian, so that any element can be factorized into irreducibles. But $6 = 2.3 = (1 + i\sqrt{5})(1 - i\sqrt{5})$, and $2, 3, 1 + i\sqrt{5}$ and $1 - i\sqrt{5}$ are irreducible in $\mathbb{Z}(i\sqrt{5})$ (why?), so that factorization is not unique. The fact that factorization need not be unique was not well understood in the beginning of the nineteenth century, leading to fallacious proofs of Fermat's last theorem.

When is factorization unique? We need a little care. Suppose that r is an element of an integral domain R that factorizes as a product $r = r_1 r_2 \cdots r_n$ of irreducible elements of R, that $\epsilon_1, \epsilon_2, \ldots, \epsilon_n$ are units in R for which $\epsilon_1 \epsilon_2 \cdots \epsilon_n = 1_R$ and that σ is a permutation of $(1, 2, \ldots, n)$. Let $r'_j = \epsilon_j r_{\sigma(j)}$

for $1 \leq j \leq n$. Then $r = r_1' r_2' \cdots r_n'$ is another factorization of r. We say that this is *essentially* the same factorization of r, and say that R is a *unique factorization domain* if every non-zero element which is not a unit can be factorized as a product of finitely many irreducible elements, and if any two factorizations are essentially the same.

If r is a non-zero element which is not a unit, in a unique factorization domain its *length* $l(r)$ is the number of terms in its factorization into irreducibles. If r is a unit, we set $l(r) = 0$. Clearly $l(rs) = l(r) + l(s)$.

In order to characterize unique factorization domains, we need a new concept. A non-zero element a of an integral domain is a *prime* if whenever $a|bc$ then either $a|b$ or $a|c$. In terms of ideals, a is a prime if and only if whenever $< bc > \subseteq < a >$ then either $< b > \subseteq < a >$ or $< c > \subseteq < a >$. A simple inductive argument shows that if a is a prime and $a|b_1 \cdots b_n$ then there exists j such that $a|b_j$.

Theorem 2.23 *A prime element of an integral domain R is irreducible.*

Proof Suppose that a is a prime, that $a|bc$, so that $a|b$ or $a|c$. Suppose that $a|b$, so that $b = af$, for some f. Then $a = afc$, so that, since R is an integral domain, $fc = 1$ and c is a unit. Similarly, if $a|c$ then b is a unit. Thus a is irreducible. $\qquad\qquad\square$

Corollary 2.24 *A prime element of $\mathbb{Z}$ is a prime number.*

Exercise

2.25 Show that a prime number is a prime element of $\mathbb{Z}$ (use the fact that $\langle p, a \rangle$ is cyclic). Deduce the fundamental theorem of arithmetic, that factorization into prime numbers is unique.

Not every irreducible element of an integral domain is prime: in $\mathbb{Z}(i\sqrt{5})$, $2|(1 + i\sqrt{5})(1 - i\sqrt{5})$, but $2 \nmid (1 + i\sqrt{5})$ and $2 \nmid (1 - i\sqrt{5})$. Primes are the key to unique factorization.

Theorem 2.25 *An integral domain R is a unique factorization domain if and only if R satisfies that $ACCPI$ condition and every irreducible element of R is a prime.*

Proof Suppose first that R is a unique factorization domain and that $(< a_i >)_{i=1}^{\infty}$ is an increasing sequence of proper principal ideals. Then the sequence $l(a_i)_{i=1}^{\infty}$ is a decreasing sequence of positive integers, and so there exists n such that $l(a_m) = l(a_n)$ for $m \geq n$. This means that a_m and a_n are associates for $m \geq n$, so that $< a_m > = < a_n >$ for $m \geq n$. Thus R satisfies the ACCPI condition. Suppose now that r is irreducible and that $r|ab$. We can write $ab = rc$. If a is a unit, $r|b$ and if b is a unit, $r|a$. Otherwise we factorize a, b and c:

$$a = s_1 \cdots s_l, \quad b = t_1 \cdots t_m, \quad c = u_1 \cdots u_n,$$

so that $s_1 \cdots s_l t_1 \cdots t_m = r u_1 \cdots u_n$. By unique factorization, r is an associate of an s_i or a t_j. Thus r divides a or b.

Suppose conversely that r satisfies the ACCPI condition and that every irreducible element of R is a prime. Then every element can be factorized into a product of irreducible elements, by Theorem 2.22. Suppose that there exist elements which do not have a unique factorization. Then among elements which do not have a unique factorization, there is an element a with a factorization $a = b_1 b_2 \cdots b_n$ into irreducible elements, with n as small as possible; a also has a different factorization $a = c_1 c_2 \cdots c_m$ into irreducible elements. Since b_1 is prime and b_1 divides $a = c_1 c_2 \cdots c_m$, there exists c_j such that $a_1|c_j$. By rearranging, we can suppose that $j = 1$, and that $c_1 = a_1 d$. Since c_1 and a_1 are irreducible, d is a unit and a_1 and c_1 are associates. Thus, if $a' = a_2 a_3 \cdots a_n$, $a' = d c_2 c_3 \cdots c_m$. But a' is the product of $n - 1$ irreducibles, and so the two products are essentially the same. Multiplying by a_1, it follows that the two original factorizations are essentially the same, giving the required contradiction.   □

If R is a unique factorization domain, factorization of elements provides an easy proof of results, as the following theorem shows.

Theorem 2.26 *Suppose that B is a non-empty subset of a unique factorization domain. Then there exists $a \in R$ such that $a|b$ for each b in B and such that if $a'|b$ for each b in B then $a'|a$.*

Such an element a is called a highest common factor of B. Any two highest common factors are associates.

Proof Pick $b_0 \in B$ and factorize it as $b_1 b_2 \cdots b_n$. If $J \subseteq \{1, \ldots, n\}$ let $p_J = \prod_{j \in J} b_j$, let $\mathcal{J} = \{J : p_J | b$ for all $b \in B\}$ and let J' be a maximal element of $\mathcal{J}$. Then $p_{J'} | b$ for all $b \in B$. If $c|b$ for all $b \in B$, then $c|b_0$, and so c is an associate of b_J for some $J \in \mathcal{J}$. Since J' is maximal, it follows that $c|p_{J'}$. Thus $p_{J'}$ is a highest common factor of B. If a and a' are highest common factors of B then $a|a'$ and $a'|a$, so that a and a' are associates.   □

If the highest common factor of B is 1_R, we say that B is *relatively prime*. If p is a highest common factor of B then $C = \{c : cp \in B\}$ is relatively prime.

Exercises

2.26 Suppose that R is an integral domain. Show that the following are equivalent:

(i) every finite non-empty set of non-zero elements of R has a highest common factor;

(ii) every finite non-empty set of non-zero elements of R has a least common multiple.

2.27 Suppose that R is an integral domain with the property that every non-empty set B of non-zero elements has a highest common factor of the form $\gamma_1 b_1 + \cdots + \gamma_n b_n$, with $b_1, \ldots, b_n$ in B and $\gamma_1, \ldots, \gamma_n$ in R. Show that R is a principal ideal domain.

2.9 Principal Ideal Domains and Euclidean Domains

Recall that a principal ideal $< a >$ in an integral domain is the set aR of all multiples of a by elements of R. An integral domain is a principal ideal domain if every ideal in R is principal. The integral domain $\mathbb{Z}$ is a principal ideal domain, since the sets $n\mathbb{Z}$ are ideals, and are the only ideals in $\mathbb{Z}$.

Theorem 2.27 *If R is a principal ideal domain, it is a unique factorization domain.*

Proof We use Theorem 2.25. R is certainly Noetherian, and so it satisfies the ACCPI condition. Suppose that a is irreducible, that $a|bc$ and that $a \nmid b$. Let $J =< a, b >$. Since $< a >$ is maximal and $b \notin < a >$, $J = R$, and there exist d, e such that $1_R = da + eb$. But then $c = dac + ebc$. Since $a|bc$, it follows that $a|c$. Thus a is prime. $\qquad\square$

In order to produce examples of principal ideal domains, we introduce another class of integral domains. A function $\phi : R \to \mathbb{Z}^+$ on an integral domain is a Euclidean function if it satisfies the following three conditions:

(i) $\phi(a) = 0$ if and only if $a = 0_R$;

(ii) if $a|b$ then $\phi(a) \le \phi(b)$;

(iii) if $a, b \in R$ and $b \neq 0$ there exist $q, r \in R$ such that $a = bq + r$, where $\phi(r) < \phi(b)$.

A Euclidean domain is an integral domain on which there exists a Euclidean function.

Theorem 2.28 *A Euclidean domain R is a principal ideal domain, and is therefore a unique factorization domain.*

Proof Let ϕ be a Euclidean function on R. Suppose that J is a non-zero ideal in R, and let $J^* = J \setminus \{0\}$. Let a be a non-zero element for which $\phi(a) = \inf\{\phi(b) : b \in J^*\}$. Suppose that $b \in J^*$. Then by (iii) there exist $q, r \in R$ such that $a = qb + r$, where $\phi(r) < \phi(a)$. But $r = a - qb \in J$, and so $r = 0$. Thus $b = qa$, and $J = <a>$. $\square$

Let us give some examples. First, $\mathbb{Z}$ is a Euclidean domain, with $\phi(n) = |n|$. Thus, as we have seen, factorization in $\mathbb{Z}$ is unique (the fundamental theorem of arithmetic). Second, let $R = \mathbb{Z}[\omega] = \mathbb{Z} + \omega\mathbb{Z}$, where $\omega = (-1 + i\sqrt{3})/2$ is a complex cube root of 1. Since $1 + \omega + \omega^2 = 0$, we can write any element r of $\mathbb{Z}(\omega)$ uniquely as $c\omega + d\omega^2$. Set $\phi(r) = |r|^2$. Since $\phi(r) = c^2 - cd + d^2$, ϕ is a non-negative integer-valued function which certainly satisfies (i) and (ii). Since the points of R form a hexagonal grid in $\mathbb{C}$ with distance 1 between adjacent points, if $a, b \in R$ and $b \neq 0$ there exists a point q in R such that $|a/b - q| < 1$. Then if $r = a - qb$, $\phi(r) < \phi(b)$. Thus $\mathbb{Z}(\omega)$ is a Euclidean domain. A similar argument shows that the ring $\mathbb{Z}(i)$ of Gaussian integers is a Euclidean domain. In contrast $\mathbb{Z}(2\omega) = \mathbb{Z}(i\sqrt{3})$ is not a Euclidean domain, since it is not a unique factorization domain ($4 = 2.2 = (1 + i\sqrt{3})(1 - i\sqrt{3})$). Similarly, $\mathbb{Z}(2i)$ is not a Euclidean domain, since it is not a unique factorization domain ($8 = 2.2.2 = (2 + 2i)(2 - 2i)$). Note that in these factorizations into irreducibles, the number of factors is different.

The next example is important enough to warrant being called a theorem.

Theorem 2.29 *If F is a field, then the integral domain $F[x]$ of polynomials over F is a Euclidean domain.*

Proof Let $\phi(a)$ be the degree of a ($\phi(0_F) = 0$). Then (i) and (ii) are satisfied, and (iii) follows by long division (an easy induction argument). Suppose that $a = a_0 + \cdots + a_n x^n$ and $b = b_0 + \cdots + b_m x^m$, where n and m are the degrees of a and b. If $m < n$, take $q = 0_F$ and if $m = n$, take $q = b_m/a_n$. Suppose that the result is true when $0 \leq m < n + k$, where $k \geq 0$. Let $q' = (b_m/a_n)x^k$ and let $b' = b - aq'$: then the degree of b' is less than $m + k$, so that there exist q'' and r so that $b' = q''a + r$, where $\phi(r) < \phi(a)$. Now take $q = q + q''$. $\square$

If a_0 and b_0 are two elements of a Euclidean domain R, we can find the highest common factor h of a_0 and b_0 without having to factorize a_0 and b_0. The method is known as *Euclid's algorithm*, and is probably known to the reader when $R = \mathbb{Z}$. We proceed as follows. Suppose that $\phi(a_0) \leq \phi(b_0)$. We choose q_0 such that if $r_0 = b_0 - q_0 a_0$, then $\phi(r_0) < \phi(a_0)$. If $r_0 = 0$ then

$a_0|b_0$, and a_0 is the highest common factor of a_0 and b_0. Otherwise, we write $\{a_1, b_1\} = \{r_0, a_0\}$ (so that $\phi(a_1) \leq \phi(b_1)$). Since $r_0 = b_0 - q_0 a_0$, $h|r_0$, so that h is the highest common factor of a_1 and b_1. But $\phi(a_1) < \phi(a_0)$, and so the process must terminate after a finite number of steps, with the first possibility.

Let us give an example. What is the highest common factor of $a_0 = 5\omega + 11\omega^2$ and $b_0 = 13\omega + 2\omega^2$ in $\mathbb{Z}(\omega)$? Since $\phi(a_0) = 91$ and $\phi(b_0) = 147$, $\phi(a_0) < \phi(b_0)$. Since the inverse of $c = a\omega + b\omega^2$ in the field of fractions $\mathbb{Q}(\omega)$ is $(b\omega + a\omega^2)/\phi(c)$, $b_0/a_0 = (-77\omega + 56\omega^2)/91$. This suggests that we take $q_0 = -\omega + \omega^2$. Then $a_0 q_0 = 17\omega + \omega^2$ and $r_0 = b_0 - a_0 q_0 = -4\omega + \omega^2$. Then $\phi(r_0) = 21$, so that we take $a_1 = r_0$ and $b_1 = a_0$. Since $b_1/a_1 = (-5\omega + 2\omega^2)/3$, we take $q_1 = -2\omega + \omega^2$. Then $q_1 a_1 = 7\omega + 14\omega^2$, so that $r_1 = b_1 - q_1 a_1 = -2\omega - 3\omega^2$. Then $\phi(r_1) = 7$, so that we take $a_2 = r_1$ and $b_2 = a_1$. But then $b_2 = (-\omega + \omega^2)a_2$, so that $r_1 = -2\omega - 3\omega^2$ is the required highest common factor.

In fact, all these messy calculations are not necessary in this case. For ϕ is a *multiplicative Euclidean function*: $\phi(ab) = \phi(a)\phi(b)$, so that h is a highest common factor of a and b if and only if $\phi(h)$ is a highest common factor of $\phi(a)$ and $\phi(b)$. Thus h is a common factor of a_0 and b_0 if and only if $\phi(h) = 7$. We have found one; a simpler associate is $\omega - 2\omega^2$.

Besides considering highest common factors, we can also consider *lowest common multiples* in a principal ideal domain.

Proposition 2.30 *Suppose that a and b are non-zero elements of a principal ideal domain R. Then there exists an element c, a lowest common multiple of a and b such that $a|c$ and $b|c$ and such that if $u|d$ and $b|d$ then $c|d$.*

Proof The set $< a > \cap < b >$ is an ideal in R, which is not 0 since $ab \in < a > \cap < b >$; thus there exists $c \in R$ such that $< c > = < a > \cap < b >$. Then $a|c$ and $b|c$, and if $a|d$ and $b|d$ then $c|d$. $\square$

A lowest common multiple is not unique; if c_1 and c_2 are lowest common multiples of a and b, then they are associates. We can obviously also define a lowest common multiple of a finite subset of R.

Exercises

2.28 Using the unique factorization domain $\mathbb{Z}[i]$, show that the equations $x^2 + 2 = y^3$ and $x^3 + 1 = y^2$ each have a unique solution in $\mathbb{N}$, which you should find.

2.29 Using the integral domain $\mathbb{Z}[i\sqrt{2}]$, show that the equation $x^2 + 2 = y^3$ has a unique solution in $\mathbb{N}$, which you should find.

2.30 Suppose that R is an integral domain. Show that $R[x]$ is a principal ideal
domain if and only if R is a field.

2.10 Polynomials Over Unique Factorization Domains

Galois theory is largely concerned with polynomials in one variable, with
coefficients in a field K. We shall, however, also need to consider polynomials
with integer coefficients, and to consider polynomials in several variables. In
order to deal with both of these, it is convenient to study polynomial rings of
the form $R[x]$, where R is a unique factorization domain.

In this section, we shall suppose that R is a unique factorization domain,
with field of fractions F. If

$$f = a_0 + a_1 x + \cdots + a_n x^n$$

is a non-zero element of $R[x]$, we define the *content* of f to be a highest
common factor of the non-zero coefficients of f (the fact that this is not
uniquely defined causes no problems). If f has content 1 we say that f is
primitive. If γ is the content of f then $f = \gamma g$, where g is primitive.

If f is an element of $R[x]$, we can consider f as an element of $F[x]$. The
next theorem provides a partial converse.

Theorem 2.31 *Suppose that R is a unique factorization domain. An element
of $R[x]$ is a unit if and only if it is a unit in R. If f is a non-zero element of
$F[x]$ we can write $f = \beta g$, where g is a primitive polynomial in $R[x]$ and
$\beta \in F$. If $f = \beta' g'$ is another such expression then g and g' are associates in
$R[x]$; there exists a unit ε in R such that $g = \varepsilon g'$.*

Proof The first statement is obvious.

Suppose that f is a non-zero element of $F[x]$. We clear denominators: there
exists δ in R such that $\delta f \in R[x]$. Let γ be the content of δf. Then $\delta f = \gamma g$,
where g is primitive in $R[x]$, and so $f = (\delta^{-1}\gamma)g = \beta g$.

Suppose that $f = \beta' g'$ is another such expression. We again clear denomi-
nators: there exists α in R such that $\alpha\beta$ and $\alpha\beta'$ are in R. Then $\alpha f = (\alpha\beta)g =
(\alpha\beta')g'$. As g is primitive in $R[x]$, $\alpha\beta$ is the content of αf: so is $\alpha\beta'$ (remember
that the content is not uniquely defined!), and so $\alpha\beta$ and $\alpha\beta'$ are associates in R.
This means that there is a unit in R such that $\alpha\beta' = \varepsilon\alpha\beta$: $\alpha\beta g = \alpha\beta' g' =
\varepsilon\alpha\beta g'$, so that $g = \varepsilon g'$ and g and g' are associates in $R[x]$. □

Theorem 2.32 *Suppose that R is a unique factorization domain. If f and g
are primitive elements of $R[x]$, so is fg.*

Proof Suppose that

$$f = a_0 + a_1 x + \cdots + a_n x^n,$$

$$g = b_0 + b_1 x + \cdots + b_m x^m,$$

and

$$fg = c_0 + c_1 x + \cdots + c_{m+n} x^{m+n}.$$

Let d be the content of fg and suppose that d is not a unit. Let r be an irreducible factor of d. As R is a unique factorization domain, r is a prime. Since f is primitive, there exists a least i such that r does not divide a_i; similarly there exists a least j such that r does not divide b_j. As r is a prime, r does not divide $a_i b_j$. We consider the coefficient

$$c_{i+j} = \sum_{k<i} a_k b_{i+j-k} + a_i b_j + \sum_{l<j} a_{i+j-l} b_l.$$

Now r divides a_k for $k < i$, and so r divides $\sum_{k<i} a_k b_{i+j-k}$; similarly r divides $\sum_{l<j} a_{i+j-l} b_l$. But r also divides c_{i+j}, and so r divides $a_i b_j$: this gives the required contradiction. □

Corollary (Gauss' lemma) *An element g of $R[x]$ is irreducible if and only if either it is an irreducible element of R or it is primitive, and irreducible in $F[x]$.*

Proof Suppose that g is irreducible in $R[x]$. If degree $g = 0$, then g must be irreducible in R. If degree $g > 0$, then g must certainly be primitive. Suppose that $g = f_1 f_2$ is a factorization in $F[x]$. By Theorem 2.31, we can write $f_1 = \beta_1 g_1, f_2 = \beta_2 g_2$ with β_1 and β_2 in F, and g_1 and g_2 primitive in $R[x]$. Thus

$$g = \beta_1 \beta_2 g_1 g_2.$$

Now $g_1 g_2$ is primitive, so that, by Theorem 2.31, $\beta_1 \beta_2$ is a unit in R. Thus $g = (\beta_1 \beta_2 g_1) g_2$, contradicting the irreducibility of g.

The converse implications are clear.

We now come to the main result of this section. □

Theorem 2.33 *If R is a unique factorization domain, so is $R[x]$*

Proof Suppose that f is a non-zero element of $R[x]$. Then $f = \alpha g$, where α is the content of f and g is primitive in $R[x]$. We now consider g as an element of $F[x]$. $F[x]$ is a unique factorization domain, and so we can write $g = g_1 \ldots g_k$ as a product of irreducible elements in $F[x]$. By Theorem 2.31, we can write each g_j as $\beta_j f_j$, where $\beta_j \in F$ and f_j is primitive in $R[x]$. Note that each f_j is irreducible in $R[x]$, by Gauss' lemma. Thus

$$g = \beta f_1 \ldots f_k,$$

where $\beta = \beta_1 \ldots \beta_k$. By Theorem 2.32, $f_1 \ldots f_k$ is primitive, and so β is a unit in R, by Theorem 2.31. Thus we can write

$$f = \alpha_1 \ldots \alpha_j f_1 \ldots f_k,$$

where $\alpha_1 \ldots \alpha_j$ is a factorization of $\alpha\beta$ as a product of irreducible elements of R. Thus f can be expressed as a product of irreducible elements of $R[x]$.

Suppose that

$$f = \alpha'_1 \ldots \alpha'_l f'_1 \ldots f'_m$$

is another such factorization. As $\alpha_1 \ldots \alpha_j$ and $\alpha'_1 \ldots \alpha'_l$ are both contents of f, they are associates in R; since R is a unique factorization domain, $l = j$ and there exists a permutation π of $\{1, \ldots, j\}$ such that α_i and $\alpha'_{\pi(i)}$ are associates for $1 \leqslant i \leqslant j$.

Further, $f_1 \ldots f_k = \lambda f'_1 \ldots f'_m$, where λ is a unit in R. By Gauss' lemma, $f_1, \ldots, f_k, f'_1, \ldots, f'_m$ are irreducible in $F[x]$. As $F[x]$ is a unique factorization domain, $m = k$ and there exists a permutation ρ of $\{1, \ldots, k\}$ and non-zero elements $\varepsilon_1, \ldots, \varepsilon_k$ of F such that $f_i = \varepsilon_i f'_{\rho(i)}$ for $1 \leqslant i \leqslant k$. But f_i and $f'_{\rho(i)}$ are primitive in $R[x]$, by Gauss' lemma, and so f_i and $f'_{\rho(i)}$ are associates in $R[x]$, by Theorem 2.31. Thus $R[x]$ is a unique factorization domain. $\square$

Corollary 1 *Suppose that f is a primitive element of $R[x]$, that g is a non-zero element of $R[x]$ and that f divides g in $F[x]$. Then f divides g in $R[x]$.*

Proof We can factorize g as

$$g = \alpha_1 \ldots \alpha_j g_1 \ldots g_k$$

where the α_i are irreducible in R and the g_i are irreducible elements of $R[x]$ of positive degree. By Gauss' lemma, each g_i is primitive and irreducible in $F[x]$. Thus

$$g = (\alpha_1 \ldots \alpha_j g_1) g_2 \ldots g_k$$

is a factorization of g as a product of irreducible elements of $F[x]$. As f divides g in $F[x]$, and as $F[x]$ is a unique factorization domain, we can write

$$f = \varepsilon g_{i_1} \ldots g_{i_r},$$

where ε is a non-zero element of F and $1 \leqslant i_1 < \ldots < i_r \leqslant k$. Now $g_{i_1} \ldots g_r$ is primitive, by Theorem 2.32. As f is also primitive, ε is a unit in R, by Theorem 2.31, and so f divides g in $R[x]$. $\qquad\square$

Corollary 2 *If R is a unique factorization domain, then so is $R[x_1, \ldots, x_n]$.*

Exercises

2.31 Express all the cubic polynomials (polynomials of degree 3) in $\mathbb{Z}_2[x]$ as products of irreducible factors.

2.32 Express all the homogeneous quadratic polynomials (polynomials of degree 2 with no constant or linear terms) in $\mathbb{Z}_2[x, y, z]$ as products of irreducible factors.

2.11 More About Fields

A field is a ring in which every non-zero element has an inverse. How can we recognize when a ring is a field?

Theorem 2.34 *A ring R is a field if and only if $\{0\}$ and R are the only ideals in R.*

Proof Suppose first that R is a field; that I is an ideal in R other than $\{0\}$ and that a is a non-zero element of I. If b is any element of R, $b = a(a^{-1}b) \in I$, and so $I = R$.

Conversely, suppose that R is a ring whose only ideals are $\{0\}$ and R. If a is a non-zero element of R, the principal ideal (a) must be R, and so there exists b in R such that $ab = 1$; consequently R is a field. $\qquad\square$

Note that if ϕ is a ring homomorphism from a field K into a ring, the kernel of ϕ is a proper ideal of K, and so ϕ is one-to-one.

Theorem 2.34 makes it easy to decide when a quotient ring is a field.

Theorem 2.35 *If J is a proper ideal in a ring R then R/J is a field if and only if J is a maximal proper ideal in R.*

Proof Let q denote the quotient map $R \to R/J$. If I is a proper ideal in R/J then $q^{-1}(I)$ is a proper ideal in R; further, $q^{-1}(I) = J$ if and only if $I = \{0\}$. It follows from this that if J is a maximal proper ideal then $\{0\}$ and R/J are the only ideals in R/J, and R/J is a field, by Theorem 2.34.

Suppose conversely that R/J is a field. If $a \notin J$, $q(a) \neq 0$, and so, since q is onto, there exists b in R such that $q(b) = (q(a))^{-1}$. Thus

$$q(ab - 1_R) = q(a)q(b) - 1_{R/J} = 0$$

so that $ab - 1_R \in J$. There therefore exists j in J such that $1_R = ab + j$, and so $(J \cup \{a\}) = R$. This means that J is a maximal proper ideal.   □

Corollary *If a is a non-zero non-unit element of a principal ideal domain R, $R/(a)$ is a field if and only if a is irreducible.*

Applying this to the ring of integers, we see that $\mathbb{Z}_n$ is a field if and only if n is a prime number.

Suppose now that K is a field. A *subfield* of K is a subset of K which is a field under the operations inherited from K. Any subfield contains 0 and 1. The intersection of all subfields is again a subfield, the smallest subfield of K. This subfield is called the *prime subfield* of K.

K is a ring; we can consider the homomorphism ϕ from $\mathbb{Z}$ into K described in Section 2.3. If K has non-zero characteristic n then, since K is certainly an integral domain, n must be a prime number. Thus $\phi(\mathbb{Z})$, which is isomorphic to $\mathbb{Z}_n$, is a field. Clearly it is the prime subfield of K.

The other possibility is that K has characteristic 0. In this case $\phi(\mathbb{Z})$ is a subring of K isomorphic to $\mathbb{Z}$. If $q = r/s$ is a rational, let us define $\phi(q)$ by setting $\phi(q) = \phi(r)\phi(s)^{-1}$. If r'/s' is another expression for q, $rs' = r's$, so that

$$\phi(r)\phi(s') = \phi(r')\phi(s)$$

and so

$$\phi(r)\phi(s)^{-1} = \phi(r')\phi(s')^{-1}.$$

Thus ϕ is properly defined, and it is equally straightforward to verify that ϕ is a ring homomorphism of $\mathbb{Q}$ into K. $\phi(\mathbb{Q})$ is a subfield of K. Clearly every element of $\phi(\mathbb{Q})$ is in every subfield of K, so that $\phi(\mathbb{Q})$ is the prime subfield of K. Summing up:

Theorem 2.36 *Suppose that K is a field. If K has characteristic 0, the prime subfield of K is isomorphic to $\mathbb{Q}$. Otherwise, K has prime characteristic, p say, and the prime subfield of K is isomorphic to $\mathbb{Z}_p$.*

2.12 Kronecker's Algorithm

Suppose that f is a polynomial in $K[x]$, where K is a field. Since $K[x]$ is a unique factorization domain, f can be expressed essentially uniquely as a product of irreducible polynomials. This raises the important practical problem: how do we recognize whether or not a given polynomial is irreducible?

There are many important cases when the field K which we consider is the field of fractions of a unique factorization domain R: this is so in the most important case of all, when the field is the field $\mathbb{Q}$ of rational numbers. In such a situation, Gauss' lemma (the corollary to Theorem 2.32) is particularly useful. Recall that Gauss' lemma implies that, if f is irreducible in $R[x]$, then f is irreducible in $K[x]$.

As an application (which we shall need later) let us consider

$$f = x^3 - 3x - 1 \in \mathbb{Z}[x].$$

As f is a cubic, if it factorized in $\mathbb{Z}[x]$ it would have a linear factor, and this would have to be either $x - 1$ or $x + 1$. But $f(1) = -3$ and $f(-1) = 1$, and so f is irreducible in $\mathbb{Z}[x]$. By Gauss' lemma, f is irreducible in $\mathbb{Q}[x]$.

In order to show the importance of Gauss' lemma, let us sketch the proof of the following result, due to Kronecker:

Theorem 2.37 *There is an algorithm to express any element of $\mathbb{Z}[x]$ as a product of irreducible factors.*

An algorithm is a procedure which takes a finite number of steps; the number of steps depends upon the polynomial in question, but an upper bound can be given for it in each case.

Proof Suppose that f has degree n. Let r be the greatest integer such that $2r \leqslant n$. If f is not irreducible, f must have a non-unit factor of degree less than or equal to r. We search for such a factor. Let $c_j = f(j)$, for $0 \leqslant j \leqslant r$. If $c_i = 0$ for some $0 \leqslant j \leqslant r$, then $x - j$ is a factor of f. Otherwise, if g is a factor of f in $\mathbb{Z}[x]$ then $g(j)$ must divide c_j for $0 \leqslant j \leqslant r$. Each c_j had finitely many divisors, and an algorithm exists to determine them. Suppose that $(d_0, \dots, d_r)$ is such that d_j is a divisor of c_j for $0 \leqslant j \leqslant r$. There exists a unique polynomial g in $\mathbb{Q}[x]$ of degree at most r such that $g(j) = d_j$ for $0 \leqslant j \leqslant r$:

$$g = \sum_{j=0}^{r} d_j g_j,$$

where

$$g_j = \prod_{0 \leqslant k \leqslant r, k \neq j} \left(\frac{x - k}{j - k} \right).$$

We can now test (by further algorithms) whether $g \in \mathbb{Z}[x]$ and whether g divides f. As there are only finitely many $(r + 1)$-tuples $(d_0, \dots, d_r)$ to consider, this means that there is an algorithm to find a non-unit factor of f, if

one exists. Repeated use of the algorithm leads to a factorization as a product
of irreducible factors.   □

Kronecker's algorithm was originally thought to be of theoretical
importance, but to be of little practical use. It can, however, be implemented by
straightforward computer programs, although now many more sophisticated
algorithms have been developed which are more suited to computer
calculations.

Exercises

2.33 Suppose that

$$f = a_0 + \cdots + a_n x^n$$

is a polynomial in $\mathbb{Z}[x]$ of degree n, and that $\max_i |a_i| = K$. Obtain
an upper bound, in terms of n and K, for the number of calculations
required to determine whether or not f is irreducible.

2.34 Suppose that K is a field with finitely many elements. Show that there is
an algorithm to express any element of $K[x]$ as a product of irreducible
factors.

2.13 Eisenstein's Criterion

It is, however, useful to have simple tests for irreducibility. Eisenstein's criterion is one such, considering the irreducibility of polynomials with coefficients
in a unique factorization domain.

Theorem 2.38 (Eisenstein's criterion) *Suppose that R is a unique factorization
domain and that $f = f_0 + f_1 x + \cdots + f_x^n \in R[x]$ is primitive. Suppose that
p is a prime in R, and that $p \mid f_i$ for $0 \leqslant i < n$, while p does not divide f_n and
p^2 does not divide f_0. Then f is irreducible in $R[x]$.*

Proof   Suppose that $f = gh$, where

$$g = g_0 + \cdots + g_r x^r$$

and

$$h = h_0 + \cdots + h_s x^s$$

are not units in R. If r were equal to 0 (so that $g = g_0$), it would follow that g_0
divides f_j for $0 \leqslant j \leqslant n$, so that g_0 would be a unit: this gives a contradiction,
so that $r \geqslant 1$. Similarly, $s \geqslant 1$. By hypothesis, p^2 does not divide $g_0 h_0$, so that

p cannot divide both g_0 and h_0. Without loss of generality we may suppose that p does not divide h_0.   □

Now $g_r h_s = f_n$, so that, by hypothesis, p does not divide g_r. Let i be the least integer such that p does not divide g_i. Then $0 \leqslant i \leqslant r < n$, so that $p \mid f_i$; that is,

$$p \mid (h_0 g_i + h_1 g_{i-1} + \cdots + h_i g_0).$$

As $p \mid g_j$ for $j < i$, $p \mid h_0 g_i$. As p is a prime, $p \mid h_0$ or $p \mid g_i$, giving a contradiction.

As an example (which we shall need later) let us observe that

$$f = x^5 - 4x + 2$$

is irreducible over $\mathbb{Z}[x]$, by Eisenstein's criterion (with $p = 2$), and so f is irreducible over $\mathbb{Q}[x]$, by Gauss' lemma.

Exercises

2.35 Suppose that R is a unique factorization domain and that

$$f = f_0 + f_1 x + \cdots + f_n x^n \in R[x]$$

has the property that $f_0, \ldots, f_n$ are relatively prime. Suppose that p is a prime in R, and that $p \mid f_i$ for $1 \leqslant i \leqslant n$, while p does not divide f_0 and p^2 does not divide f_n. Show that f is irreducible in $R[x]$.

2.36 Show that if p is a prime number then $x^n - p$ is irreducible in $\mathbb{Q}[x]$.

2.37 Show that $x^5 - 4x + 2$ and $x^4 - 4x + 2$ are irreducible over $\mathbb{Q}(i)$.

2.14 Localization

Even if Eisenstein's criterion cannot be applied directly, it is sometimes possible to apply it after making a suitable transformation. For example, if

$$f = x^4 + 4x^3 + 10x^2 + 12x + 7 \in \mathbb{Z}[x],$$

it is not possible to apply Eisenstein's criterion directly. If we write $y = x + 1$, we find that

$$f = y^4 + 4y^2 + 2.$$

As

$$g = x^4 + 4x^2 + 2$$

is irreducible in $\mathbb{Z}[x]$, by Eisenstein's criterion, f must be irreducible too. The problem of course is to find a suitable transformation: this is a matter of ingenuity and good fortune.

Exercises

2.38 Show (by making the transformation $y = x - 1$) that if p is a prime number then $1 + x + \cdots + x^{p-1}$ is irreducible over $\mathbb{Q}$.

2.39 Let $\theta = 2\pi/7$. What is the minimal polynomial of $e^{i\theta}$ over $\mathbb{Q}$? What is the minimal polynomial of $2\cos\theta$ over $\mathbb{Q}$?

2.40 How many elements are there in the set S of quintic polynomials in $\mathbb{Z}_2[x]$? How many have a linear factor? Show that there is one irreducible quadratic polynomial in $\mathbb{Z}_2[x]$. How many irreducible polynomials are there in S? Determine them explicitly.

2.41 Carry out a similar analysis for the set of quintic polynomials in $\mathbb{Z}_3[x]$. Does it give more information about the irreducible quintic polynomials in $\mathbb{Z}[x]$?

There is another technique which can sometimes prove useful when we are considering polynomials in $\mathbb{Z}[x]$. Suppose that p is a prime number: for each integer n, let $\bar{n}$ denote the image (mod p) of n under the quotient map from $\mathbb{Z}$ to $\mathbb{Z}_p$. This quotient map induces a ring homomorphism from $\mathbb{Z}[x]$ onto $\mathbb{Z}_p[x]$; if

$$f = a_0 + a_1 x + \cdots + a_n x^n \in \mathbb{Z}[x],$$

then

$$\bar{f} = \bar{a}_0 + \bar{a}_1 x + \cdots + \bar{a}_n x^n \in \mathbb{Z}_p[x].$$

Theorem 2.39 (Localization principle) *Suppose that*

$$f = a_0 + a_1 x + \cdots + a_n x^n \in \mathbb{Z}[x],$$

and that $a_0, \ldots, a_n$ are relatively prime. Suppose that p is a prime which does not divide a_n. If $\bar{f}$ is irreducible in $\mathbb{Z}_p[x]$, then f is irreducible in $\mathbb{Z}[x]$.

Proof Suppose that f factors as $f = gh$, where g and h are not units. As in the proof of Eisenstein's criterion, since $a_0, \ldots, a_n$ are relatively prime, degree $g \geqslant 1$ and degree $h \geqslant 1$; of course degree $g +$ degree $h =$ degree f. $\qquad\square$

As p does not divide a_n, degree $\bar{f} =$ degree f. As $\bar{f} = \bar{g}\bar{h}$, degree $\bar{f} =$ degree $\bar{g} +$ degree $\bar{h}$. As degree $\bar{g} \leqslant$ degree g, and degree $\bar{h} \leqslant$ degree h, we must have that degree $\bar{g} =$ degree $g \geqslant 1$ and degree $\bar{h} =$ degree $h \geqslant 1$. Thus $\bar{f} = \bar{g}\bar{h}$ is a non-trivial factorization of $\bar{f}$.

Notice that the localization principle can also be used to establish Eisen-stein's criterion in $\mathbb{Z}[x]$. With the notation of Theorem 2.38, $\overline{f} = \overline{f}_n x^n$ (why?). Consequently, as $\overline{f} = \overline{g}\overline{h}$, $g_0 = 0 \pmod{p}$ and $h_0 = 0 \pmod{p}$ so that $f_0 = g_0 h_0 = 0 \pmod{p^2}$, giving a contradiction.

To give another example of the use of localization, let us show that $f = x^n + px + p^2$ is irreducible in $\mathbb{Z}[x]$ (where p is a prime number). First observe that if α is a root of f in $\mathbb{Z}$, then $\alpha < 0$ and $\alpha = 0 \pmod{p}$ so that $\alpha = -kp$ for some positive integer k. From this it follows that

$$(-k)^n p^{n-2} = k - 1,$$

which clearly has no solution. Thus, if $f = gh$ is a factorization in $\mathbb{Z}[x]$, degree $g \geqslant 2$ and degree $h \geqslant 2$. Arguing as before, $g_i = 0 \pmod{p}$ for $i < $ degree g and $h_j = 0 \pmod{p}$ for $j < $ degree h, so that

$$p = g_0 h_1 + g_1 h_0 = 0 \pmod{p^2}$$

giving a contradiction.

3

Vector Spaces and Determinants

3.1 Vector Spaces

The reader will probably be familiar with the notion of a real or complex vector space. In Galois theory we need to consider vector spaces over an arbitrary field K: we replace the field of scalars $\mathbb{R}$ or $\mathbb{C}$ by an arbitrary field K. The details are very similar, although special results hold when K is finite.

Suppose that K is a field. A set V is a *vector space over* K if first it is an abelian group under addition and second there is a mapping $(\alpha, x) \to \alpha x$ from $K \times V$ into V which satisfies

(a) $\alpha(x + y) = \alpha x + \alpha y$,
(b) $(\alpha + \beta)x = \alpha x + \beta x$,
(c) $(\alpha\beta)x = \alpha(\beta x)$, and
(d) $1 . x = x$

for all α, β in K and x, y in V.

As an example, let S be a non-empty set, and let K^s denote the set of all mappings from S into K. If f and g are in K^s, define $f + g$ by

$$(f + g)(s) = f(s) + g(s), \text{ for } s \text{ in } S,$$

and, if $\alpha \in K$, define αf by

$$(\alpha f)(s) = \alpha(f(s)), \text{ for } s \text{ in } S.$$

Then it is easy to verify that the axioms are satisfied. If $S = \{1, \ldots, n\}$, we write K^n for K^s and, if $x \in K^n$, write

$$x = (x_1, \ldots, x_n),$$

where x_j is the value of x at j.

54

Another example that we shall study in detail is the following. Suppose that L is a field, and that K is a subfield. Then L is a vector space over K, when the algebraic operations are defined in the obvious way. For example, $\mathbb{C}$ is a vector space over $\mathbb{R}$ and $\mathbb{R}$ is a vector space over $\mathbb{Q}$. A subset W of a vector space V over K is a *linear subspace* if, with the same operations of addition and multiplication, it is a vector space over K.

If V_1 and V_2 are vector spaces over the same field K, then a {K-linear mapping} (or simply a linear mapping, when the context is clear) from V_1 to V_2 is a mapping $T : V_1 \to V_2$ which satisfies

$$T(a+b) = T(a) + T(b), \quad T(\lambda a) = \lambda T(a)$$

for all $a, b \in V_1$ and all $\lambda \in K$. The set of linear mappings from V_1 to V_2 is denoted by $L(V_1, V_2)$: if $V_1 = V_2$ it is also denoted by $L(V_1)$. When the operations are defined in the obvious way, $L(V_1, V_2)$ is also a vector space over K: if $T \in L(V_1, V_2)$ then $T^{-1}(0)$ is the null-space of T, and $T(V_1)$ the image of T.

Vector spaces lend themselves to elementary linear geometry. If V is a vector space over K then the line $l(a, b)$ is the set

$$l(a, b) = \{(1 - \lambda)a + \lambda b : \lambda \in K\} = a + \{\theta(b - a) : \theta \in K\}.$$

Exercise

3.1   Suppose that V is a vector space over K, that W is a linear subspace of V and that $l(a, b)$ is a line in V. Then either $l(a, b) \subset W$ or $l(a, b) \cap W$ has at most one point.

Theorem 3.1 *Suppose that $W_1, \ldots, W_n$ are proper linear subspaces of a vector space V over an infinite field K. Then $V \neq \cup_{j=1}^{n} W_j$.*

Proof   The proof is by induction. The result is certainly true when $n = 1$. Suppose that it is true for $n - 1$. Let a be an element of V not in $\cup_{j=1}^{n-1} W_j$ and let b be an element not in W_n. If $a = b$ there is nothing to prove. Otherwise, $l(a, b)$ meets each W_j in at most one point. Since $l(a, b)$ has infinitely many points, there is a point c of $l(a, b)$ not in any W_j.   □

In fact, we need surprisingly little of the theory of vector spaces. The key is the idea of *dimension*; as we shall see, this turns out to be remarkably powerful. Suppose that V is a vector space over K. A subset W of V is a *linear subspace* if it is a vector space under the operations defined on V, for this, it is sufficient that if x and y are in W and α is in K then $x+y \in W$ and $\alpha x \in W$. If A is a non-empty subset of V, the *span* of A, denoted by span (A), is the intersection of the linear subspaces containing A; it is a linear subspace of V, and is the smallest linear subspace containing A. If span $(A) = V$, we say that A *spans V*.

We now turn to linear dependence and linear independence. A subset A of a vector space V over K is *linearly dependent over K* if there are finitely many distinct elements $x_1, \ldots, x_k$ of A and elements $\lambda_1, \ldots, \lambda_k$ of K, *not all zero*, such that

$$\lambda_1 x_1 + \cdots + \lambda_k x_k = 0;$$

if A is not linearly dependent over K, A is *linearly independent over K*. Note that, even if A is infinite, the sums which we consider are finite. If A is finite and $A = \{x_1, \ldots, x_n\}$, *where the x_i are distinct*, A is linearly independent over K if it follows from

$$\lambda_1 x_1 + \cdots + \lambda_n x_n = 0$$

that $\lambda_1 = \lambda_2 = \cdots = \lambda_n = 0$.

A subset A of a vector space V over K is a *basis* for V if it is linearly independent and spans V.

Exercise

3.2   Suppose that V is a vector space over K. If A is a non-empty subset of V, show that the span of A is the set

$$\left\{ b \in V : b = \sum_{n=1}^{m} \lambda_n a_n, \lambda_n \in K, a_n \in A, m \in \mathbb{N} \right\}.$$

Our main interest is in finite-dimensional vector spaces; let us consider them now.

A vector space V over K is *finite dimensional* if there exists a finite subset of V which spans V. First we show that a finite-dimensional space has a finite basis; this is a consequence of the following theorem.

Theorem 3.2 *Suppose that A is a finite subset of a vector space V over K which spans V, and that C is a linearly independent subset of A (C may be empty). There exists a basis B of V with $C \subseteq B \subseteq A$.*

Proof   Consider the collection J of all subsets of A which contain C and are linearly independent. Since $|A| < \infty$, there exists a B in J with a maximum number of elements. B is independent and $C \subseteq B \subseteq A$; it remains to show that B spans V.

Let $B = \{b_1, \ldots, b_n\}$, where the b_i are distinct. If $a \in A \backslash B$, $B \cup \{a\}$ is linearly dependent (by the maximality of $|B|$) and so there exist $\lambda_0, \ldots, \lambda_n$ in K, not all zero, such that

$$\lambda_0 a + \lambda_1 b_1 + \cdots + \lambda_n b_n = 0.$$

Further, $\lambda_0 \neq 0$, for otherwise $b_1, \ldots, b_n$ would be linearly dependent. Thus

$$a = -\lambda_0^{-1} \lambda_1 b_1 - \lambda_0^{-1} \lambda_2 b_2 - \cdots - \lambda_0^{-1} \lambda_n b_n$$

and $a \in \text{span}(B)$. Consequently $A \subseteq \text{span}(B)$, and so $\text{span}(A) \subseteq \text{span}(B)$. As $\text{span}(A) = V$, the theorem is proved. $\square$

We would now like to define the dimension of a finite-dimensional vector space as the number of elements in a basis. In order to do this, we need to show that any two bases have the same number of elements. This follows from the next theorem.

Theorem 3.3 *Suppose that V is a vector space over K. If A spans V and C is a linearly independent subset of V, then $|C| \leqslant |A|$.*

Proof The result is trivially true if $|A| = \infty$, and so we may suppose that $|A| < \infty$. If $|C| = \infty$, there is a finite subset D of C with $|D| > |A|$. As D is again linearly independent, it is sufficient to prove the result when $|C| < \infty$. $\square$

Theorem 3.3 is therefore a consequence of the following.

Theorem 3.4 (The Steinitz exchange theorem) *Suppose that $C = \{c_1, \ldots, c_r\}$ is a linearly independent subset (with r distinct elements) of a vector space V over K, and that $A = \{a_1, \ldots, a_s\}$ is a set (with s distinct elements) which spans V. Then there exists a set D, with $C \subseteq D \subseteq A \cup C$, such that $|D| = s$ and D spans V.*

Proof We prove this by induction on r. The result is trivially true for $r = 0$ (take $D = A$). Suppose that it is true for $r - 1$. As the set $C_0 = \{c_1, \ldots, c_{r-1}\}$ is linearly independent, there exists a set D_0 with $C_0 \subseteq D_0 \subseteq A \cup C_0$ such that $|D_0| = s$, and D_0 spans V. By relabelling A if necessary, we can suppose that

$$D_0 = \{c_1, \ldots, c_{r-1}, a_r, a_{r+1}, \ldots, a_s\}.$$

If s were equal to $r - 1$, we would have $D_0 = C_0$; but $c_r \in \text{span}(D_0)$, so that we could write

$$c_r = \sum_{i=1}^{r-1} \gamma_i c_i,$$

contradicting the linear independence of C. Thus $s \geqslant r$. As $c_r \in \mathrm{span}\,(D_0)$, we can write

$$c_r = \sum_{i=1}^{r-1} \gamma_l c_i + \sum_{j=r}^{s} \alpha_j a_j.$$

Not all α_j can be zero, for again this would contradict the linear independence of C. By relabelling if necessary, we can suppose that $\alpha_r \neq 0$. Let $D = \{c_1, \ldots, c_r, a_{r+1}, \ldots, a_s\}$. Then

$$a_r = \alpha_r^{-1} \left(c_r - \sum_{i=1}^{r-1} \gamma_i c_i - \sum_{j=r+1}^{s} \alpha_j a_j \right)$$

so that $a_r \in \mathrm{span}\,(D)$. Thus

$$\mathrm{span}\,(D) \supseteq \{c_1, \ldots, c_{r-1}, a_r, a_{r+1}, \ldots, a_s\} = D_0$$

and so $\mathrm{span}\,(D) \supseteq \mathrm{span}\,(D_0) = V$.

This completes the proof. $\qquad\qquad\qquad\qquad\qquad\qquad\qquad\qquad\square$

Corollary (to Theorem 3.3) *Any two bases of a finite-dimensional vector space have the same finite number of elements.*

We now define the *dimension* of a finite-dimensional vector space V over K to be the number of elements in a basis. We denote the dimension of V by $\dim V$. If V is not finite dimensional over K we set $\dim V = \infty$. Here is one simple but important result:

Theorem 3.5 *Suppose that U is a linear subspace of a finite-dimensional vector space V over K. Then $\dim U \leqslant \dim V$, and $\dim U = \dim V$ if and only if $U = V$.*

Proof Let A be a basis for U, and let C be a finite set which spans V. Considered as a subset of V, A is linearly independent, and so by Theorem 3.2 there is a basis B of V with $A \subseteq B \subseteq A \cup C$. Thus

$$\dim U = |A| \leqslant |B| = \dim V.$$

If $\dim U = \dim V$, we must have $A = B$, so that A spans V and $U = V$; of course if $U = V$, $\dim U = \dim V$. $\qquad\qquad\qquad\qquad\qquad\qquad\square$

Corollary 1 *Suppose that A is a finite subset of a finite-dimensional vector space V over K. If $|A| > \dim V$, A is linearly dependent.*

Proof Let $U = \text{span}(A)$. If A were linearly independent, A would be a basis for U, so that $\dim U = |A|$. But $\dim U \leqslant \dim V$, giving a contradiction.   □

Suppose that V_1 and V_2 are vector spaces over the same field K. A mapping ϕ from V_1 into V_2 is called a *linear mapping* if

$$\phi(x + y) = \phi(x) + \phi(y),$$
$$\phi(\lambda x) = \lambda \phi(x)$$

for all x and y in V_1 and all λ in K. The study of linear mappings is an essential part of the study of vector spaces; for our purposes we shall only need one further corollary to Theorem 3.5.

Corollary 2 *Suppose that V_1 and V_2 are vector spaces over K and that ϕ is a linear mapping of V_1 into V_2. If $\dim V_1 > \dim V_2$, ϕ is not one-to-one, and there exists a non-zero x in V_1 such that $\phi(x) = 0$.*

Proof Let $n = \dim V_2$. As $\dim V_1 > \dim V_2$, there exist $n + 1$ linearly independent vectors $x_1, \ldots, x_{n+1}$ in V_1. Then, by Corollary 1, $\{\phi(x_1), \ldots, \phi(x_{n+1})\}$ is linearly dependent in V_2, and so there exist $\lambda_1, \ldots, \lambda_{n+1}$ in K, not all zero, such that

$$\lambda_1 \phi(x_1) + \cdots + \lambda_{n+1} \phi(x_{n+1}) = 0.$$

But

$$\lambda_1 \phi(x_1) + \cdots + \lambda_{n+1} \phi(x_{n+1}) = \phi(\lambda_1 x_1 + \cdots + \lambda_{n+1} x_{n+1}),$$

since ϕ is linear, and

$$x = \lambda_1 x_1 + \cdots + \lambda_{n+1} x_{n+1} \neq 0$$

since $\{x_1, \ldots, x_{n+1}\}$ is linearly independent. As $\phi(x) = 0 = \phi(0)$, ϕ is not one-to-one.   □

Suppose that V_1 and V_2 are finite-dimensional vector spaces over the same field and that $T \in L(V_1, V_2)$. The *image* $T(V_1)$ is a linear subspace of V_2; its dimension is the *rank* $r(T)$ of T. Its null space $N(T) = T^{-1}(0)$ is a linear subspace of V_1; its dimension is the *nullity* $n(T)$ of T.

Theorem 3.6 *Suppose that $T \in L(V_1, V_2)$, where V_1 is finite dimensional. Then $n(T) + r(T) = \dim(V_1)$.*

Proof Let $(a_1, \ldots, a_n)$ be a basis for $n(T)$. Extend it to a basis $(a_1, \ldots, a_n, b_1, \ldots, b_j)$ of V_1. Then $T(b_1), \ldots, T(b_j)$ span $T(v_1)$. But they are also linearly independent, for if $\lambda_1 T(b_1) + \cdots + \lambda_j T(b_j) = 0$, then $T(\lambda_1 b_1 + \cdots + \lambda_j b_j) = 0$, and $(\lambda_1 b_1 + \cdots + \lambda_j b_j) \in N(T)$. Since $b_1, \ldots, b_j$ are linearly independent, $\lambda_i = 0$ for $1 \le i \le j$; thus $b_1, \ldots, b_j$ are linearly independent, and therefore form a basis for $T(V_1)$. Hence $r(T) = j$. Since $\dim(V_1) = n + j$, the result follows. $\qquad\qquad\qquad\qquad\qquad\qquad\qquad\qquad\qquad\qquad\qquad\qquad\quad\square$

Corollary 3.7 *Suppose that V_1 and V_2 are finite-dimensional vector spaces over K.*

(i) *There is a linear isomorphism of V_1 onto V_2 if and only if V_1 and V_2 have the same dimension.*
(ii) *If $T \in L(V_1, V_2)$ and $\dim(V_1) > \dim(V_2)$ then there exists $x \ne 0$ such that $T(x) = 0$.*

Exercises

3.3   In K^n, let $e_j = (0, \ldots, 0, 1, 0, \ldots, 0)$, where the 1 occurs in the jth position. Let $f_j = e_1 + \cdots + e_j$.

(a) Show that $\{e_1, \ldots, e_n\}$ is a basis for K^n.
(b) Show that $\{f_1, \ldots, f_n\}$ is a basis for K^n.
(c) Is $\{e_1, f_1, f_2, \ldots, f_n\}$ a basis for K^n?

3.4   Suppose that S is infinite. For each s in S, let $e_s(t) = 1$ if $s = t$, and let $e_s(t) = 0$ otherwise. Is $\{e_s : s \in S\}$ a basis for K^s?

3.5 $\mathbb{R}$ can be considered as a vector space over $\mathbb{Q}$. Show that $\mathbb{R}$ is *not* finite dimensional over $\mathbb{Q}$. Can you find an infinite subset of $\mathbb{R}$ which is linearly independent over $\mathbb{Q}$?

3.6   Suppose that K is an infinite field and that V is a vector space over K. Show that it is not possible to write $V = \bigcup_{i=1}^n U_i$, where $U_1, \ldots, U_n$ are proper linear subspaces of V.

3.7   Suppose that V is a finite-dimensional vector space and that $T \in L(V)$. Show that $n(T^j) \le jn(T)$.

3.2 The Infinite-Dimensional Case

Suppose that V is an infinite-dimensional space. Does it have a basis? We begin with the countable case.

Theorem 3.8 *Suppose that V is a vector space with countably many points. Then V has a basis.*

Proof   Suppose that $V = (x_n)_{n=1}^{\infty}$. Let $B = \{x_n : x_n \notin \mathrm{span}\{x_j : j < n\}\}$. Then B is linearly independent, and spans V. $\qquad\qquad\qquad\qquad\qquad\quad\square$

If V is uncountable, we need Zorn's lemma (see the Appendix).

Theorem 3.9 *Suppose that A is a subset of a vector space V over K which spans V and that C is a linearly independent subset of A (C may be empty). There exists a basis B of V with $C \subseteq B \subseteq A$.*

We have proved this in the case that A is finite in Theorem 3.2 (and made essential use of the finiteness of A). Taking $A = V$ and C the empty set, we see that this theorem implies that every vector space has a basis.

Proof Let S denote the collection of subsets of A which are linearly independent and contain C. Order S by inclusion. Suppose that T is a chain in S. Let $E = \bigcup_{D \in T} D$. E is a subset of A which contains C. Suppose that $x_1, \ldots, x_n$ are distinct elements of E. From the definition of E, there are sets $D_1, \ldots, D_n$ in T such that $x_i \in D_i$ for $1 \leqslant i \leqslant n$. Since T is a chain, there exists j, with $1 \leqslant j \leqslant n$, such that $D_i \subseteq D_j$ for $1 \leqslant t \leqslant n$. Consequently $x_1, \ldots, x_n$ are all in D_j. As D_j is linearly independent, $\{x_1, \ldots, x_n\}$ is linearly independent. As this holds for any finite subset of E, E is linearly independent. Thus $E \in S$. E is clearly an upper bound for T, and so every chain in S has an upper bound. $\qquad\square$

We can therefore apply Zorn's lemma, and conclude that S has a maximal element B. B is linearly independent and $C \subseteq B \subseteq A$; it remains to show that span $(B) = V$. Since span$(A) = V$, it is enough to show that if $a \in A \setminus B$ then $a \in$ span(B). Let $B_0 = \{a\} \cup B$. Then B_0 is not linearly independent, and so there exist $b_1, \ldots, b_n$ distinct elements of B and $\lambda_0, \lambda_1, \ldots, \lambda_n$, not all 0, in K, such that $\lambda_0 a + \lambda_1 b_1 + \cdots, \lambda_n b_n = 0$; since $b_1, \ldots, b_n$ are linearly independent, $\lambda_0 \neq 0$ and $a = \lambda_0^{-1}(\lambda_1 b_1 + \cdots + \lambda_n b_n) \in$ span(B).

3.3 Characters and Automorphisms

Let us now establish results about linear independence that we shall need later.

Suppose that G is a group and that K is a field. A (*K-valued*) *character* on G is a homomorphism of G into the multiplicative group K^* of non-zero elements of K. We can think of a character as a K-valued function on G; recall that the set of all K-valued functions on G is a vector space over K.

Theorem 3.10 *Suppose that G is a group, that K is a field and that S is a set of K-valued characters on G. Then S is linearly independent over K.*

Proof If not, there is a minimal non-empty subset $\{\gamma_1, \ldots, \gamma_n\}$ of distinct elements of S which is linearly dependent over K. That is, there exist non-zero $\lambda_1, \ldots, \lambda_n$ in K such that

$$\lambda_1 \gamma_1(g) + \cdots + \lambda_n \gamma_n(g) = 0 \qquad (*)$$

for all g in G. Each γ_i is non-zero, since it sends the identity of G to 1, and so $n \geqslant 2$. As $\gamma_1 \neq \gamma_n$, there exists h in G such that $\gamma_1(h) \neq \gamma_n(h)$. Now

$$\lambda_1 \gamma_1(hg) + \cdots + \lambda_n \gamma_n(hg) = 0$$

for all g in G. Using the fact that the γ_i are characters, we have that

$$\lambda_1 \gamma_1(h)\gamma_1(g) + \cdots + \lambda_n \gamma_n(h)\gamma_n(g) = 0$$

for all g in G. Now multiply $(*)$ by $\gamma_n(h)$ and subtract:

$$\lambda_1(\gamma_1(h) - \gamma_n(h))\gamma_1(g) + \cdots + \lambda_{n-1}(\gamma_{n-1}(h) - \gamma_n(h))\gamma_{n-1}(g) = 0$$

for all g in G. As $\gamma_1(h) - \gamma_n(h) \neq 0$, this means that $\{\gamma_1, \ldots, \gamma_{n-1}\}$ is linearly dependent over K, contradicting the minimality of $\{\gamma_1, \ldots, \gamma_n\}$. $\square$

If τ is an automorphism of a field K, then the restriction of τ to K^* is a K-valued character on K^*. Spelling the theorem out in detail in this case, we have the following corollary:

Corollary *Suppose that $\tau_1, \ldots, \tau_n$ are distinct automorphisms of a field K and that $k_1, \ldots, k_n$ are non-zero elements of K. Then there exists k in K such that*

$$k_1 \tau_1(k) + \cdots + k_n \tau_n(k) \neq 0.$$

Corollary 3.11 (Dedekind's lemma) *Suppose that S is a set of field homomorphisms from a field F_1 into a field F_2. Then S is a linearly independent set of F_2-valued functions on F_1.*

Proof For the restrictions of S to the multiplicative group F_1^* form a linearly independent set of F_2-valued functions on F_1^*. $\square$

3.4 Determinants

The theory of determinants is very large, and we shall only consider the basic results that we shall need.

Suppose that K is a field and that $n \in \mathbb{N}$. A K-valued $n \times n$ matrix A is a function from $n \times n$ to K; we write $A(ij)$ as a_{ij}. The ith row r_i of A is the element $(a_{ij})_{j=1^n}$ of K^n, and the jth column c_j is the element $(a_{ij})_{i=1}^n$ of K^n.

This defines a linear mapping from K^n into itself: if $x = (x_1, \ldots, x_n)$ then $A(x) = y$, where $y_i = \sum_{j=1}^{n} a_{ij}x_j$. For example, let $I_n(ij) = 1$ if $i = j$ and let $I_n(ij) = 0$ otherwise. I_n is the unit matrix: it defines the identity mapping on K^n.

We use the determinant of A to determine when A is invertible.

Let $M_n(K)$ be the set of $n \times n$ matrices; it is a vector space over K of dimension n^2. If $A \in M_n(K)$, we define the determinant of A to be

$$\det(A) = \det(c_1, \ldots, c_n) = \sum_{\sigma \in \Sigma_n} \epsilon_\sigma a_{1\sigma(1)} \cdots a_{n\sigma(n)},$$

where ϵ_σ is the signature of σ.

The determinant has the following properties:

Theorem 3.12 *Suppose that K is a field and that $A = (a_{ij}) \in K^n$: let $A = (c_1 \ldots c_n)$, where c_j is the jth column of A.*

(i) *If $\lambda \in K$ then $\det(\lambda c_1, c_2, \ldots, c_n) = \lambda \det(A)$.*

(ii) *$\det(c_1, c_2, \ldots, c_n) + \det(c_1', c_2, \ldots, c_n) = \det(c_1 + c_1', c_2, \ldots, c_n)$.*

(iii) *If $\tau \in \Sigma_n$ then $\det(c_{\tau(1)}, \ldots, c_{\tau(n)}) = \epsilon_\tau \det(c_1, \ldots, c_n)$.*

(iv) *$\det(c_1, \ldots, c_n) = \det(r_1, \ldots, r_n)$, where r_i is the ith row of A.*

(v) *$\det(c_1, \ldots, c_n) = 0$ if and only if $c_1, \ldots, c_n$ are linearly dependent.*

Proof (i), (ii) and (iii) follow easily from the definition.

(iv)

$$\det(r_1, \ldots, r_n) = \sum_{\sigma \in \Sigma_n} \epsilon_\sigma a_{\sigma(1)1} \cdots a_{\sigma(n)n}$$

$$= \sum_{\sigma \in \Sigma_n} \epsilon_\sigma a_{1\sigma^{-1}(1)} \cdots a_{n\sigma^{-1}(n)}$$

$$= \sum_{\sigma \in \Sigma_n} \epsilon_\sigma^{-1} a_{1\sigma(1)} \cdots a_{n\sigma(n)}$$

$$= \sum_{\sigma \in \Sigma_n} \epsilon_\sigma a_{1\sigma(1)} \cdots a_{n\sigma(n)}$$

$$= \det(c_1, \ldots, c_n).$$

(v) Suppose first that $a_1, \ldots, a_n$ are linearly dependent. By renumbering if necessary, there exist $\lambda_1 = 1, \lambda_2, \ldots, \lambda_n$ such that $b_1 = \sum_{i=1}^{n} \lambda_i a_i = 0$. But then $\det(a_1, \ldots, a_n) = \det(b_1, a_2, \ldots, a_n) = 0$.

If $a_1, \ldots, a_n$ are linearly independent, then, by rearranging if necessary and using (i)–(iii), we can construct a triangular matrix $(t_1, \ldots, t_n)$ such that $t_{ii} \neq 0$

for $1 \le i \le n$ and $t_{ij} = 0$ for $j < i$, and such that $\det(\mathbf{t}_1, \ldots, \mathbf{t}_n) = \det(\mathbf{a}_1, \ldots, \mathbf{a}_n)$. But $\det(\mathbf{t}_1, \ldots, \mathbf{t}_n) = \prod_{i=1}^{n} t_{nn} \ne 0$. □

Theorem 3.13 *If $A, B \in M_n(K)$ then $\det(AB) = \det(A).\det(B)$.*

Proof Since $A\mathbf{b}_j = \sum_{i=1}^{n} a_i b_{ij}$,

$$\det(AB) = \det(A\mathbf{b}_1, \ldots, A\mathbf{b}_n)$$

$$= \det\left(\sum_{i_1=1}^{n} \mathbf{a}_{i_1} b_{i_1 1}, \ldots, \sum_{j_n=1}^{n} \mathbf{a}_{i_n} b_{i_n n}\right)$$

$$= \sum_{i_1 \ldots i_n} \det(\mathbf{a}_1, \ldots, \mathbf{a}_n) \epsilon (i_1, \ldots, i_n) b_{i_1 1}, \ldots, b_{i_n n}$$

$$= \det(A).\det(B).$$ □

Exercise

3.8 Use (iv) and (v) of Theorem 3.4 repeatedly to show that if $A \in M_n(K)$ there exists $T \in M_n(K)$ for which $\det(T) = \det(A)$ and $T_{ij} = 0$ for $i > j$ (T is *triangular*). What is $\det(T)$? Use this result to give another proof that if $A, B \in M_n(K)$ then $\det(AB) = \det(A)\det(B)$.

This means that we can define the determinant of an endomorphism of a finite-dimensional space. Suppose that V is finite dimensional and that $S \in L(V, V)$. Let j and j' be linear isomorphisms of K^n onto V.

Theorem 3.14 $\det(j^{-1}Sj) = \det(j'^{-1}Sj')$.

Proof Let $L = j^{-1}j'$. Then $\det(j^{-1}Sj) = \det(L^{-1}(j^{-1}Sj)L) = \det(j'^{-1}Sj')$. □

We can therefore define $\det(S) = \det(j^{-1}Sj)$.

Theorem 3.15 *Suppose that $A \in L(V)$, where V is a finite-dimensional vector space over K. Then A has an inverse A^{-1} ($AA^{-1} = A^{-1}A = I$) if and only if $\det(A) \ne 0$.*

Proof Let $j : K^n \to V$ be a linear isomorphism, so that A has an inverse if and only if $B = j^{-1}Aj$ does, and $\det(A) = \det(B)$. If $B = (\mathbf{b}_1, \ldots, \mathbf{b}_n)$, then B ihas an inverse if and only if $(\mathbf{b}_1, \ldots, \mathbf{b}_n)$ are linearly independent, which happens if and only if $\det(B) \ne 0$. □

Exercises

3.9 Suppose that $d : M_n(K) \to K$ is linear in each variable $\mathbf{a}_j$, that $d(\mathbf{e}_1, \ldots, \mathbf{e}_n) = 1$ and that $d(\mathbf{a}_1, \ldots, \mathbf{a}_n) = 0$ if $\mathbf{a}_i = \mathbf{a}_j$ for some $i \neq j$. Show that $d = \det$.

3.10 Suppose that $A, B, C, D \in M_n(K)$. Define $E \in M_{2n}(K)$ and show that $det(E) = \det(A)\det(D) - \det(B)\det(C)$.

3.11 Suppose that $A, B \in M_n(K)$ and that A is invertible. Show that $A + \lambda B$ is invertible for all but finitely many λ in K.

Suppose that $A \in M_n(K)$, that $x \in K^n$, that $\lambda \in K$ and that $Ax = \lambda x$. Then x is an eigenvector of A and λ an eigenvalue. The polynomial $p(x) = \det(A - xI)$ is the characteristic polynomial of A.

3.12 Suppose that A and B are non-zero elements of $M_n(K)$ and that $AB = 0$. Show that $\det(A) = \det(B) = 0$.

Suppose that $A, B, C \in M_n(K)$ and that $BC = I_n$. Show that A and AB have the same characteristic polynomial.

3.13 Suppose that $A, P \in M_n(K)$ and that $P^2 = P$. Show that AP and PA have the same characteristic polynomial.

3.14 Suppose that $A, B \in M_n(K)$. Show that there is an invertible $J \in M_n(K)$ such that JA is a projection. Deduce that AB and BA have the same characteristic polynomial.

3.15 Suppose that $A \in M_n(K)$. How would you find the eigenvalues and eigenvectors of A? How many eigenvalues can there be for which the equation $A(x) = \lambda x$ has a solution in K^n?

3.16 Suppose that $(b_j)_{j=1}^n \in K^n$. Define $A \in M_n(A)$ by setting $a_{i,i+1} = 1$, for $1 \leq i < n$, $a_{n,j} = b_{n-j+1}$ and $a_{ij} = 0$ for other values of i and j. What is the characteristic polynomial of A? What polynomials in $K[x]$ can be characteristic polynomials?

3.17 Let $\phi = (\phi_1, \phi_2)$ be a bijection from $(1, \ldots, n_1 n_2)$ onto $(1, \ldots, n_1) \times (1, \ldots, n_2)$. Suppose that $A \in M_{n_1}(K)$ and $B \in M_{n_2}(K)$. Let $C \in M_{n_1 n_2}$ be defined as $c_{ij} = a_{\phi_1(i), \phi_1(j)} b_{\phi_2(i), \phi_2(j)}$. Show that $\det(C) = \det(A)^{n_1} \det(B)^{n_2}$.

PART II

The Theory of Fields and Galois Theory

4

Field Extensions

4.1 Introduction

One of the main topics of Galois theory is the study of polynomial equations. In order to consider how we should proceed, let us first consider some rather trivial and familiar examples.

Polynomials involve addition and multiplication, and so it is natural to consider polynomials with coefficients in a ring R. If we consider the simplest possible case, when $R = \mathbb{Z}$ and p is a polynomial of degree 1, we find there are difficulties: for example, we cannot solve the equation $2x + 3 = 0$ in $\mathbb{Z}$.

In the case where R is an integral domain, the field of fractions is constructed in order to deal with this problem. Thus, in the example above, if we consider 2 and 3 as elements of $\mathbb{Q}$, the rational field, then the equation has a solution $x = -3/2$ in $\mathbb{Q}$.

Let us now consider a quadratic equation: $x^2 - 2x - 1 = 0$. We consider this as an equation with rational coefficients: completing the square, we find that

$$(x - 1)^2 = 2.$$

But there is no rational number r for which $r^2 = 2$. For if $r = a/b$ in the lowest terms, then $a^2 = 2b^2$, so that a is even, and $a = 2c$, say. But then $b^2 = 2c^2$, so that b is also even, giving a contradiction.

Instead, the first natural idea is to consider the polynomial as a polynomial with *real* coefficients: the equation then factorizes as $(x - 1 + \sqrt{2})(x - 1 - \sqrt{2}) = 0$, and we have solutions $1 - \sqrt{2}$ and $1 + \sqrt{2}$.

The field $\mathbb{R}$ is rather large, however ($\mathbb{R}$ is uncountable, while $\mathbb{Q}$ is countable), and it is possible to proceed more economically. The set of all real numbers of the form $a + b\sqrt{2}$, where a and b are rational, forms a field K: addition and multiplication are obvious, and $(a + b\sqrt{2})^{-1} = c + d\sqrt{2}$ where $c = a/(a^2 - 2b^2)$ and $d = -b/(a^2 - 2b^2\sqrt{2})$. Clearly $\mathbb{Q} \subseteq K \subseteq \mathbb{R}$, and

K is much smaller than $\mathbb{R}$, since K is countable and $\mathbb{R}$ is not. If we consider $x^2 - 2x - 1$ as an element of $K[x]$, we can solve the equation in K.

Let us express all this in more algebraic language. The polynomial $x^2 - 2x - 1$ is irreducible in $\mathbb{Q}[x]$ (and is therefore irreducible in $\mathbb{Z}[x]$). If, however, it is considered as an element of $K[x]$ or $\mathbb{R}[x]$, it can be written as a product of linear factors. This suggests the following general programme: given an element f in $K[x]$ (where K is a field), can we find a larger field, L say, such that f considered as an element of $L[x]$ can be written as a product of linear factors? If so, can we do it in an economical way?

4.2 Field Extensions

Suppose that we start with a field K. In order to construct a larger field L we frequently have, by some means or another, to construct L, and then find a subfield of L which is isomorphic to K (think of how the complex numbers are constructed from the reals). It is occasionally important to realize that this sort of procedure is adopted: for this reason we define an *extension* of a field K to be a triple (i, K, L), where L is another field, and i is a (ring) monomorphism of K into L.

However, much more frequently this is far too cumbersome. If (i, K, L) is an extension of K, the image $i(K)$ is a subfield of L which is isomorphic to K; we shall usually identify K with $i(K)$ and consider it as a subfield of L. In this case we shall write $L : K$ for the extension. Thus $\mathbb{C} : \mathbb{R}$ is the extension of the real numbers by the complex numbers and $\mathbb{R} : \mathbb{Q}$ is the extension of the rational numbers by the real numbers. Very occasionally, when the going gets rough, we shall need to be rather careful: in these circumstances we shall revert to the notation (i, K, L).

Suppose now that $L : K$ is an extension. How do we measure how big the extension is? It turns out that the appropriate idea is dimension, in the vector space sense. This is an almost embarrassingly simple idea: the remarkable thing is that it is extraordinarily powerful.

To begin with, then, we forget about many of the field properties of L.

Theorem 4.1 *Suppose that $L : K$ is an extension. Under the operations*

$$(l_1, l_2) \to l_1 + l_2 \text{ from } L \times L \text{ to } L$$

and

$$(k, l) \to kl \text{ from } K \times L \text{ to } L,$$

L is a vector space over K.

Proof   All the axioms are satisfied.   □

Thus $\mathbb{C}$ is a real vector space, and $\mathbb{R}$ is a vector space over the rationals $\mathbb{Q}$.

We now define the *degree of an extension* $L : K$ to be the dimension of L as a vector space over K. We write $[L : K]$ for the degree of $L : K$. We say that $L : K$ is *finite* if $[L : K] < \infty$, and that $L : K$ is *infinite* if $[L : K] = \infty$.

Thus $[\mathbb{C} : \mathbb{R}] = 2$, $[\mathbb{R} : \mathbb{Q}] = \infty$, and, if K is the field of all $r + s\sqrt{2}$, with r and s rational, $[K : \mathbb{Q}] = 2$. In this sense, then, $K : \mathbb{Q}$ is a more economical extension for solving $x^2 - 2x - 1 = 0$ than $\mathbb{R} : \mathbb{Q}$.

The next theorem is very straightforward (there is an obvious argument to try, and it works), but it is the key to much that follows. If $M : L$ and $L : K$ are extensions, then clearly so is $M : K$.

Theorem 4.2 *Suppose that $M : L$ and $L : K$ are extensions. Then*

$$[M : K] = [M : L][L : K].$$

Proof   First suppose that the right-hand side is finite, so that we can write $[M : L] = m < \infty$, and $[L : K] = n < \infty$. Let $(x_1, \ldots, x_m)$ be a basis for M over L, and let $(y_1, \ldots, y_n)$ be a basis for L over K. We can form the products $y_j x_i$ (for $1 \leqslant i \leqslant m$, $1 \leqslant j \leqslant n$) in M. We shall show that the mn elements $(y_j x_i : 1 \leqslant i \leqslant m, 1 \leqslant j \leqslant n)$ form a basis for M over K.

First we show that they span M over K. Let $z \in M$. As $(x_1, \ldots, x_m)$ is a basis for M over L, there exist $\alpha_1, \ldots, \alpha_m$ in L such that

$$z = \alpha_1 x_1 + \cdots + \alpha_m x_m.$$

As each α_i is in L, and as $(y_1, \ldots, y_n)$ is a basis for L over K, for each i there exist $\beta_{i1}, \ldots, \beta_{in}$ in K such that

$$\alpha_i = \beta_{i1} y_1 + \cdots + \beta_{in} y_n.$$

Substituting,

$$z = \sum_{i=1}^{m} \sum_{j=1}^{n} \beta_{ij} y_j x_i,$$

which proves our assertion.

Second we show that $(y_j x_i : 1 \leqslant i \leqslant m; 1 \leqslant j \leqslant n)$ is a linearly independent set over K. Suppose that

$$0 = \sum_{i=1}^{m} \sum_{j=1}^{n} \gamma_{ij} y_j x_i$$

where the γ_{ij} are elements of K. Let us set

$$\delta_i = \sum_{j=1}^{n} \gamma_{ij} y_j (\in L)$$

for $1 \leqslant i \leqslant m$. Then

$$0 = \sum_{i=1}^{m} \delta_i x_i.$$

But $(x_1, \ldots, x_m)$ is linearly independent over L, and so $\delta_i = 0$ for $1 \leqslant i \leqslant m$; that is,

$$0 = \sum_{j=1}^{n} \gamma_{ij} y_j, \text{ for } 1 \leqslant i \leqslant m.$$

Now $\gamma_{ij} \in K$, and $(y_1, \ldots, y_n)$ is a linearly independent set over K. Consequently $\gamma_{ij} = 0$ for all i and j, and the second assertion is proved. Thus the elements $(y_j x_i)_i \overset{m}{=} 1, j \overset{n}{=} 1$ form a basis for M over K, and

$$[M : K] = [M : L][L : K]$$

provided that the right-hand side is finite.

If $[M : K] = l < \infty$, we can find a basis $(z_1, \ldots, z_l)$ for M over K. $(z_1, \ldots, z_l)$ spans M over K, and so it certainly spans M over L. Thus $[M : L] < \infty$. Also L is a K-linear subspace of M, so that $[L : K] < \infty$. Thus, if the right-hand side is infinite, we must have $[M : K] = \infty$: the proof is complete.   □

We can extend this result in an obvious way. A sequence $K_n : K_{n-1}, K_{n-1} : K_{n-2}, \ldots, K_1 : K_0$ of extensions, where each field extends its successor, is called a *tower*. Clearly

$$[K_n : K_0] = [K_n : K_{n-1}][K_{n-1} : K_{n-2}] \ldots [K_1 : K_0];$$

we refer to this (and to Theorem 4.2) as the *tower law* for field extensions.

Exercises

4.1   Suppose that $[L : K]$ is a prime number. What fields are there intermediate between L and K?

4.2   Suppose that $[L : K] = 2$ and that the characteristic of K is not 2. Show that there exists $\alpha \in L$ such that $\alpha^2 \in K$ and $L = K(\alpha)$.

4.3 Construct addition and multiplication tables for all fields with four elements. Does the result of the preceding question hold for fields of characteristic 2?

4.4 Suppose that $[L : K]$ is odd, and that $\alpha \in L$. Show that $L(\alpha) = L(\alpha^2)$.

4.5 Suppose that $[K(\alpha) : K] = m$ and $[K(\beta) : K] = n$, where m and n are coprime. Show that $[K(\alpha, \beta) : K] = mn$. Suppose that f is an irreducible polynomial in $K[x]$ of degree m. Show that it has no roots in $K(\beta)$.

4.6 Suppose that $r(x) = p(x)/q(x)$ is a non-constant rational function in $K(x)$. Find a polynomial f in $K(r)[y]$ for which $f(x) = 0$.

4.3 Algebraic and Transcendental Extensions

Suppose that $L : K$ is an extension, and that A is a subset of L. We write $K(A)$ for the intersection of all subfields of L which contain K and A. $K(A)$ is a subfield of L, and is the smallest subfield of L containing K and A. Clearly $L : K(A)$ and $K(A) : K$ are extensions. $K(A) : K$ is the *extension of K generated by A*.

It is useful to see what a typical element of $K(A)$ looks like. Let

$$S = \{\alpha_1 \ldots \alpha_k : \alpha_i \in A \cup \{1\}\}$$

be the set of all finite products of elements of A, together with 1, let V be the K-linear subspace of L generated by S and let $V^* = V \setminus \{0\}$. Then

$$K(A) = \{rs^{-1} : r \in V, s \in V^*\};$$

for clearly anything on the right-hand side belongs to $K(A)$, and it is a straightforward matter to verify that the right-hand side is a subfield of L containing K and A.

If $A = \{\alpha_1, \ldots, \alpha_n\}$ is finite, we say that $K(A) : K$ is a *finitely generated extension*, and write $K(\alpha_1, \ldots, \alpha_n)$ for $K(A)$. In particular, we say that an extension $L : K$ is *simple* if there exists α in L such that $L = K(\alpha)$. Thus $\mathbb{C} : \mathbb{R}$ is a simple extension, since $\mathbb{C} = \mathbb{R}(i)$. Similarly, the field K of all $m + n\sqrt{2}$, with m and n in $\mathbb{Q}$, is $\mathbb{Q}(\sqrt{2})$.

It follows from the description of $K(A)$ that, if $L : K$ is a simple extension of K and if K is countable, then L is also countable; thus $\mathbb{R} : \mathbb{Q}$ is not a simple extension.

Suppose now that $L : K$ is an extension and that $\alpha \in L$. There are two possibilities. First, there may be a non-zero polynomial $f = k_0 + k_1 x + \cdots + k_n x^n$ in $K[x]$ such that

$$f(\alpha) = k_0 + k_1 \alpha + \cdots + k_n \alpha^n = 0.$$

In other words, α is a *root* of f. In this case we say that α is *algebraic* over K. Second, it may happen that no such polynomial exists: in this case we say that α is *transcendental* over K. The two possibilities lead to very different developments: for the time being we shall concentrate on algebraic elements, and consider transcendental elements at a much later stage (Chapter 19).

At this point, let us remark that the study of transcendental numbers – that is, elements of $\mathbb{R}$ or $\mathbb{C}$ which are transcendental over $\mathbb{Q}$ – is one of the most difficult and profound areas of number theory. It was not until 1844 that Liouville showed that any transcendental numbers exist: this helps us to understand why Cantor's set theory, which shows that there are uncountably many transcendental numbers (see Exercise 4.17), came as such a shock. Cantor's result is of no help in particular cases: Hermite's result that e is transcendental was proved in 1873, the year before Cantor's result, and the fact that π is transcendental was proved by Lindemann in 1882. The proofs are analytical, and far away from the material of this book. For an account of transcendental number theory, see the book by Baker.[1]

Let us express these ideas in terms of mappings. Suppose that $L : K$ is an extension and that $\alpha \in L$. We define the *evaluation map* E_α from $K[x]$ into L by setting $E_\alpha(f) = f(\alpha)$ for each f in $K[x]$. Notice that E_α is a ring homomorphism from $K[x]$ into L. It then follows immediately from the definitions that α is transcendental over K if and only if E_α is injective and that α is algebraic over K if and only if E_α is not injective.

Suppose that α is algebraic over K. The kernel K_α of the evaluation map E_α is a non-zero ideal in $K[x]$; as $K[x]$ is a principal ideal domain, there is a non-zero polynomial m_α such that $K_\alpha = (m_\alpha)$. Further, since the non-zero elements of K are the units in $K[x]$, we can take m_α to be *monic* (that is, m_α has leading coefficient 1:

$$m_\alpha = k_0 + k_1 x + \cdots + k_{n-1} x^{n-1} + x^n,$$

and then m_α is uniquely determined. The polynomial m_α is called the *minimal polynomial* of α.

---

[1] A. Baker, *Transcendental Number Theory*, Cambridge University Press, 1979.

Theorem 4.3 *Suppose that $L : K$ is an extension and that $\alpha \in L$ is algebraic. Then m_α is irreducible in $K[x]$, the image $E_\alpha(K[x])$ of the polynomial ring $K[x]$ is the subfield $K(\alpha)$ of L and we can factorize E_α as $i\tilde{E}_\alpha q$:*

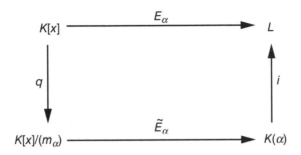

where q is the quotient mapping, $\tilde{E}_\alpha$ is an isomorphism and i is the inclusion mapping.

Proof Suppose that $m_\alpha = fg$. Then

$$0 = E_\alpha(m_\alpha) = E_\alpha(f)E_\alpha(g) = f(\alpha)g(\alpha),$$

so that either $f(\alpha) = 0$ or $g(\alpha) = 0$. If $f(\alpha) = 0$, $f \in (m_\alpha)$, so that $m_\alpha|f$ and g is a unit. Similarly, if $g \in (m_\alpha)$, f is a unit. Thus m_α is irreducible. The corollary to Theorem 2.35 implies that $K[x]/(m_\alpha)$ is a field. Now by Theorem 2.1 we can factorize E_α in the following way:

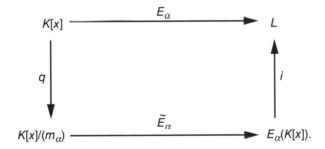

Since $\tilde{E}_\alpha$ is an isomorphism, this means that $E_\alpha(K[x])$ is a subfield of L. Since $E_\alpha(k) = k$ if $k \in K$ and $E_\alpha(x) = \alpha$, $E_\alpha(K[x]) \supseteq K \cup \{\alpha\}$, and so $E_\alpha(K[x]) \supseteq K(\alpha)$. But clearly $E_\alpha(K[x]) \subseteq K(\alpha)$, and so the proof is complete.   □

Corollary 4.4 *If $f \in K[x]$, then $f(\alpha) = 0$ if and only if $m_\alpha|f$.*

Let us now relate these ideas to the degree of an extension.

Theorem 4.5 *Suppose that $L : K$ is an extension and that $\alpha \in L$. Then α is algebraic over K if and only if $[K(\alpha) : K] < \infty$. If this is so, then $[K(\alpha) : K]$ is the degree of m_α.*

Proof First, suppose that $[K(\alpha) : K] = n < \infty$. Consider the $n + 1$ terms 1, $\alpha, \alpha^2, \ldots, \alpha^n$ in $K(\alpha)$. Either two terms α^r and α^z (with $0 \leqslant r < s \leqslant n$) are equal, in which case $x^s - x^r$ is in the kernel K_α of the evaluation map E_α, or they are all distinct. In this latter case, by Corollary 4.4 to Theorem 3.5, $\{1, \alpha, \ldots, \alpha^n\}$ are linearly dependent over K. Thus there exist $k_0, k_1, \ldots, k_n$, not all zero, such that $k_0 + k_1\alpha + \cdots + k_n\alpha^n = 0$. Then

$$f = k_0 + k_1 x + \cdots + k_n x^n \in K_\alpha$$

so that in either case E_α is not one-to-one.

Next suppose that α is algebraic over K, and that m_α is the minimal polynomial of α. We shall show that if $n = $ degree (m_α) then $\{1, \alpha, \ldots, \alpha^{n-1}\}$ forms a basis for $K(\alpha)$ over K. First we show that $\{1, \alpha, \ldots, \alpha^{n-1}\}$ is a linearly independent set over K. For if

$$k_0.1 + k_1\alpha + \cdots + k_{n-1}\alpha^{n-1} = 0,$$

let us set $f = k_0 + k_1 x + \cdots + k_{n-1}x^{n-1}$. Then $f \in K_\alpha = (m_\alpha)$ and degree $f < $ degree m_α, so that $f = 0$, and $k_0 = k_1 = \cdots = k_{n-1} = 0$. Second we show that $\{1, \alpha, \ldots, \alpha^{n-1}\}$ spans $K(\alpha)$. By Theorem 4.3, if $\beta \in K(\alpha)$ then $\beta = E_\alpha(f)$ for some $f \in K[x]$. We can write

$$f = m_\alpha q + r$$

where $r = 0$ or degree $r < n$. Then $\beta = E_\alpha(f) = E_\alpha(m_\alpha)E_\alpha(q) + E_\alpha(r) = E_\alpha(r)$ so that if $r = k_0 + k_1 x + \cdots + k_{n-1}x^{n-1}$,

$$\beta = k_0 + k_1\alpha + \cdots + k_{n-1}\alpha^{n-1} \in \text{span}(1, \alpha, \ldots, \alpha^{n-1}). \qquad \square$$

Exercises

4.7 Suppose that $L : K$ and that K_1 and K_2 are two intermediate fields such that $L = K(K_1, K_2)$. Show that $[L : K] \leqslant [K_1 : K][K_2 : K]$.

4.8 Suppose that $K(\alpha) : K$ is a finite simple extension. For each β in $K(\alpha)$, let $T_\alpha(\beta) = \alpha\beta$. T_α is a linear mapping of $K(\alpha)$ (considered as a vector space over K) into itself. Show that $\det(xI - T_\alpha)$ is the minimal polynomial of α over K.

4.9 Show that $x^3 + 3x + 1$ is irreducible in $\mathbb{Q}[x]$. Suppose that α is a root of $x^3 + 3x + 1$ in $\mathbb{C}$. Express α^{-1} and $(1 + \alpha)^{-1}$ as linear combinations, with rational coefficients, of $1, \alpha$ and α^2.

4.10 Suppose that $L(\alpha) : L : K$ and that $[K(\alpha) : K]$ and $[L : K]$ are relatively prime. Show that the minimal polynomial of α over L has its coefficients in K.

4.11 Suppose that $[L : K]$ is a prime number. Show that $L : K$ is simple.

4.12 Suppose that $L : K$ and that $\alpha, \beta \in L$ are transcendental over K. Show that α is algebraic over $K(\beta)$ if and only if β is algebraic over $K(\alpha)$.

4.13 Suppose that $\alpha = \sqrt{p} + \sqrt{q}$, where p and q are distinct prime numbers. Show that α is algebraic over $\mathbb{Q}$. What is the minimal polynomial for α? What is $[\mathbb{Q}\alpha : \mathbb{Q}]$? Find a basis for $\mathbb{Q}(\alpha)$ as a vector space over $\mathbb{Q}$.

4.14 Suppose that $L : K$ is an extension and that α, β are elements of L for which $\alpha + \beta$ and $\alpha\beta$ are algebraic over K. Show that α and β are algebraic over K.

4.15 Let $w = a + ib \in \mathbb{C}$, let $w^2 = c + id$, and suppose that $|w| = 1$ and that $2(a + c) = -1$. Show that a is algebraic over $\mathbb{Q}$, and find its minimal polynomial. Show that b is algebraic over $\mathbb{Q}$, and find its minimal polynomial. What is $[\mathbb{Q}(b) : \mathbb{Q}]$? Calculate w^3, and show that w is algebraic over $\mathbb{Q}$. What is its minimal polynomial over $\mathbb{Q}$? What is $[\mathbb{Q}(w) : \mathbb{Q}]$? Find a basis for $\mathbb{Q}(w)$.

4.16 Let $\mathbb{Z}_3$ be the field with three elements. Find all the irreducible monic quadratic polynomials over $\mathbb{Z}_3$. How many non-isomorphic extensions $L : \mathbb{Z}_3$ of dimension 2 are there?

4.4 Algebraic Extensions

Theorem 4.5 has the following important consequence.

Theorem 4.6 *Suppose that $L : K$ is an extension. The set L_a of those elements of L which are algebraic over K is a subfield of L.*

Proof Suppose that α and β are in L_a. As β is algebraic over K, β is certainly algebraic over $K(\alpha)$. As $K(\alpha)(\beta) = K(\alpha, \beta)$, we have $[K(\alpha, \beta) : K(\alpha)] < \infty$, by Theorem 4.5. Also $[K(\alpha) : K] < \infty$, by Theorem 4.5, so that, by Theorem 4.2,

$$[K(\alpha, \beta) : K] = [K(\alpha, \beta) : K(\alpha)][K(\alpha) : K] < \infty.$$

Now $K(\alpha + \beta) \subseteq K(\alpha, \beta)$, and so $[K(\alpha + \beta) : K] < \infty$. Using Theorem 4.5 again, we see that $\alpha + \beta$ is algebraic over K. Similarly, $\alpha\beta$ is algebraic over K. Finally, if α is a non-zero element of L_a, with minimal polynomial

$$f = k_0 + k_1 x + \cdots + k_{n-1} x^{n-1} + x^n$$

let $g = 1 + k_{n-1}x + \cdots + k_0 x^n$. Then $g(\alpha^{-1}) = \alpha^{-n} f(\alpha) = 0$ so that α^{-1} is algebraic over K.

This theorem gives some indication of how useful the idea of the degree of an extension is. We have shown that if α and β are algebraic over K then so are $\alpha + \beta$ and $\alpha\beta$, but we have not had to produce polynomials in $K[x]$ of which these elements are roots.

We say that an extension $L : K$ is *algebraic* if every element of L is algebraic over K. Not every algebraic extension is finite: for example, if A denotes the *algebraic numbers*, the set of complex numbers which are algebraic over $\mathbb{Q}$, then $A : \mathbb{Q}$ is infinite (see Exercise 4.23 below). Finite extensions are characterized in the following way: □

Theorem 4.7 *Suppose that $L : K$ is an extension. The following are equivalent:*

(i) $[L : K] < \infty$;

(ii) *$L : K$ is algebraic, and L is finitely generated over K;*

(iii) *there exist finitely many algebraic elements $\alpha_1, \ldots, \alpha_n$ of L such that $L = K(\alpha_1, \ldots, \alpha_n)$.*

Proof Suppose first that $L : K$ is finite. If $\alpha \in L$, then $[K(\alpha) : K] \leqslant [L : K] < \infty$, so that α is algebraic over K (Theorem 4.5); thus $L : K$ is algebraic. If $(\beta_1, \ldots, \beta_r)$ is a basis for L over K, then $L = K(\beta_1, \ldots, \beta_r)$, so that L is finitely generated over K. Thus (i) implies (ii), and (ii) trivially implies (iii).

Suppose now that (iii) holds. Let $K_0 = K$, and let $K_j = K(\alpha_1, \ldots, \alpha_j) = K_{j-1}(\alpha_j)$, for $1 \leqslant j \leqslant n$. Note that $L = K_n$. Each α_j is algebraic over K_{j-1}, so that $[K_j : K_{j-1}] < \infty$. We have a tower of extensions, and consequently

$$[L : K] = [K_n : K_0] = [K_n : K_{n-1}][K_{n-1} : K_{n-2}]\ldots[K_1 : K_0] < \infty.$$

□

Corollary 1 *If $L : K$ is an extension and if α is an element of L which is algebraic over K, then $K(\alpha) : K$ is algebraic.*

This is a special case of the next corollary.

Corollary 2 *Suppose that $L : K$ is an extension and that $S \subset L$. If each $\alpha \in S$ is algebraic over K, then $K(S) : K$ is algebraic.*

Proof If $\beta \in K(S)$, there exist $\alpha_1, \ldots, \alpha_n$ in S such that $\beta \in K(\alpha_1, \ldots, \alpha_n)$. By the theorem, $K(\alpha_1, \ldots, \alpha_n) : K$ is algebraic, and so β is algebraic over K.   □

The proof of this corollary shows that, even though an algebraic extension may be infinite, it is possible to deal with it by using arguments involving finite extensions. The same is true of the next result.

Theorem 4.8 *Suppose that* $M : L$ *and* $L : K$ *are algebraic extensions. Then* $M : K$ *is algebraic.*

Proof Suppose that $\alpha \in M$, and that

$$m_\alpha = l_0 + l_1 x + \cdots + l_n x^n$$

is its minimal polynomial over L. (Since m_α is monic, $l_n = 1$.) Then α is algebraic over $K(l_0, \ldots, l_n)$ and so

$$[K(l_0, \ldots, l_n)(\alpha) : K(l_0, \ldots, l_n)] = [K(l_0, \ldots, l_n, \alpha) : K(l_0, \ldots, l_n)] < \infty$$

by Theorem 4.5. Also

$$[K(l_0, \ldots, l_n) : K] < \infty$$

by Theorem 4.7, and so

$$[K(\alpha) : K] \leqslant [K(l_0, \ldots, l_n, \alpha) : K]$$
$$= [K(l_0, \ldots, l_n, \alpha) : K(l_0, \ldots, l_n)][K(l_0, \ldots, l_n) : K]$$
$$< \infty;$$

thus α is algebraic over K. $\qquad\qquad\qquad\qquad\qquad\qquad \square$

Exercises

4.17 Show that if $L : K$ is algebraic and K is countable then L is countable. Show that there exist real numbers which are transcendental over the rationals.

4.18 Suppose that $L : K$ is an extension, that α is an element of L which is transcendental over K, and that f is a non-constant element of $K[x]$. Show that $f(\alpha)$ is transcendental over K. Show that, if β is an element of L which satisfies $f(\beta) = \alpha$, then β is transcendental over K.

4.19 Suppose that a and b are complex numbers which are transcendental over $\mathbb{Q}$. Is a^b transcendental over $\mathbb{Q}$?

4.20 Suppose that $K(\alpha, \beta) : K$ is an extension, that α is algebraic over K, but not in K, and that β is transcendental over K. Show that $K(\alpha, \beta) : K$ is not simple.

4.5 Monomorphisms of Algebraic Extensions

The next result uses finiteness in a rather different way. If $L : K$ is an extension and $\tau : L \to L$ is a monomorphism with the property that $\tau(k) = k$ for each k in K, we say that τ *fixes* K.

Theorem 4.9 *Suppose that $L : K$ is algebraic and that $\tau : L \to L$ is a ring monomorphism which fixes K. Then τ maps L onto L.*

Proof   Certainly $\tau(0) = 0$. Suppose that α is a non-zero element of L. Let m_α be its minimal polynomial over K. Let R be the set of roots of m_α in L. If $\beta \in R$,

$$m_\alpha(\tau(\beta)) = \tau(m_\alpha(\beta)) = \tau(0) = 0$$

so that τ maps R into R. Now τ is one-to-one and R is finite, and so τ must map R onto R. Thus there exists β in R such that $\tau(\beta) = \alpha$. As this holds for each α in L, τ must map L onto L.

A ring monomorphism of a field onto itself is called an *automorphism*.   □

Exercises

4.21   Show that the condition that $L : K$ is algebraic cannot be dropped from Theorem 4.9.

4.22   Let A denote the field of real numbers which are algebraic over $\mathbb{Q}$. Show that $[A : \mathbb{Q}] = \infty$.

4.23   Show that the positive pth roots of 2 (as p varies over the primes) are linearly independent over $\mathbb{Q}$.

5

Ruler and Compass Constructions

5.1 Some Classical Problems

Many of the problems that exercised Greek mathematicians and their successors were geometric, and in particular concerned constructions using ruler (straight edge) and compasses. Here are the three most important.

(i) *Squaring the circle*. Construct a square whose arc is the same as a circle of given radius.

(ii) *Duplicating the cube*. Construct a length l such that a cube with side l has twice the volume of a given cube.

(iii) *Trisecting the angle $\pi/3$*. It is easy to construct two lines that meet at an angle $\alpha = \pi/3$. Trisect this angle: construct two lines that meet at an angle $\alpha/3$.

The Greeks laboured in vain: none of them is possible. For the first, we need to know Lindemann's result that π is transcendental, but otherwise the results follow simply from the ideas of Chapter 4. It is remarkable that these ideas, which are really rather elementary, resolve the issues decisively: an idea does not need to be complicated to be effective.

5.2 Constructible Points

There are many constructions that one can carry out with ruler (straight edge) and compasses alone.

We are given two points in the plane, which we take as 0 and 1, and call p_0 and p_1. We call the line containing them l_1 and call c_1 the circle with centre p_0 and radius $[p_0, p_1]$. Having constructed $(p_0, \ldots, p_n)$, $(l_0, \ldots, l_n)$ and $(c_0, \ldots, c_n)$, we then call p_{n+1} the intersection of two lines or circles previously constructed, and construct by ruler and compasses the line l_{n+1}

containing two points previously constructed and the circle $n + 1$ with a centre point previously constructed and a radius the distance between two previously constructed points. What points can we construct in this way?

Exercises

5.1   Following this procedure, how would you construct the following?

    (a) The line through 0 perpendicular to the real axis l_1.

    (b) The line through a constructed point perpendicular to a constructed line.

    (c) The line through a constructed point parallel to a constructed line.

    (d) n equally spaced points between two constructed points on a given line.

    This means that we can construct any point $(x, y) \in \mathbb{Q} \times \mathbb{Q}$.

5.2   (a) Show that if two constructed lines l and l' meet at a point p, then we can construct a line bisecting the angle between them.

    (b) Construct $\sqrt{2}$, $\sqrt{n}$ (use induction) and $\sqrt{r}$, where r is rational.

Can we characterize the points that can be constructed using ruler and compasses? We say that such a point is *constructible*, and denote the set of constructible points by S.

We now wish to associate some fields to these geometric ideas. We do this in a very straightforward way: $\mathbb{R} : \mathbb{Q}$ is an extension; if $P = (x, y)$ is a constructible point, we consider the extension $\mathbb{Q}(x, y) : \mathbb{Q}$ generated by x and y.

Theorem 5.1 *If $P = (x, y)$ is a constructible point, the extension $\mathbb{Q}(x, y) : \mathbb{Q}$ is finite, and $[\mathbb{Q}(x, y) : \mathbb{Q}] = 2^r$, for some non-negative integer r.*

Proof   Since P is constructible, there exists a sequence $P_0, P_1, \ldots, P_n = P$ of points which satisfies the requirements of the definitions. Let $P_j = (x_j, y_j)$, and for $1 \leqslant j \leqslant n$ let

$$F_j = \mathbb{Q}(x_1, y_1, x_2, y_2, \ldots, x_j, y_j).$$

Then $F_{j+1} = F_j(x_{j+1}, y_{j+1})$, for $1 \leqslant j < n$. We shall show that $[F_{j+1} : F_j] = 1$ or 2: then, by the tower law, $[F_n : F_1] = [F_n : \mathbb{Q}] = 2^s$ for some non-negative integer s. But $\mathbb{Q}(x, y) = \mathbb{Q}(x_n, y_n)$ is a subfield of F_n containing $\mathbb{Q}$, so that, by the tower law again,

$$[F_n : \mathbb{Q}(x, y)][\mathbb{Q}(x, y) : \mathbb{Q}] = 2^s,$$

and so $[\mathbb{Q}(x, y) : \mathbb{Q}] = 2^r$, for some non-negative integer r.

It remains to show that $[F_{j+1} : F_j] = 1$ or 2.

If (a_1, b_1) and (a_2, b_2) are two points in S_j, the equation of the line joining (a_1, b_1) and (a_2, b_2) is $(x - a_2)(b_1 - b_2) = (a_1 - a_2)(y - b_2)$, and therefore has the form

$$\lambda x + \mu y + v = 0,$$

where λ, μ and v are elements of F_j. Similarly the equation of the circle, centre (a_1, b_1) and radius the distance between points (a_2, b_2) and (a_3, b_3) of S, is

$$(x - a_1)^2 + (y - b_1)^2 = (a_2 - a_3)^2 + (b_2 - b_3)^2,$$

and therefore has the form

$$x^2 + y^2 + 2gx + 2fy + c = 0,$$

where f, g and c are elements of F_j.

We are now in a position to consider the three cases that can arise.

Case (i). (x_{j+1}, y_{j+1}) is the intersection of two distinct straight lines, each joining two points of S. In this case (x_{j+1}, y_{j+1}) is the solution of two simultaneous equations

$$\lambda_1 x + \mu_1 y + v_1 = 0,$$
$$\lambda_2 x + \mu_2 y + v_2 = 0$$

with coefficients in F_j. Solving these, we find that x_{j+1} and y_{j+1} are in F_j, so that $F_{j+1} = F_j$ and $[F_{j+1} : F_j] = 1$.

Case (ii). (x_{j+1}, y_{j+1}) is a point of intersection of an appropriate straight line and circle. In this case (x_{j+1}, y_{j+1}) satisfies equations

$$\lambda x + \mu y + v = 0,$$
$$x^2 + y^2 + 2gx + 2fy + c = 0$$

with coefficients in F_j. Suppose that $\lambda \neq 0$. We can then eliminate x, and obtain a monic quadratic equation in y. If this factors over F_j as

$$(y - \alpha)(y - \beta) = 0$$

then $y_{j+1} = \alpha$ or β, so that $y_{j+1} \in F_j$; substituting in the linear equation, $x_{j+1} \in F_j$, so that $F_{j+1} = F_j$, and $[F_{j+1} : F_j] = 1$. If the quadratic is irreducible, it must be the minimal polynomial for y_{j+1}: thus, by Theorem 4.5, $[F_j(y_{j+1}) : F_j] = 2$. As $x_{j+1} = -\lambda^{-1}(\mu y_{j+1} + v), x_{j+1} \in F_j(y_{j+1})$ and so $F_{j+1} = F_j(x_{j+1}, y_{j+1}) = F_j(y_{j+1})$. If $\lambda = 0$, then $\mu \neq 0$, and we can repeat the argument, interchanging the roles of x_{j+1} and y_{j+1}.

Case (iii). $(x_{j+1},\ y_{j+1})$ is a point of intersection of two suitable circles. In this case $(x_{j+1},\ y_{j+1})$ satisfies equations

$$x^2 + y^2 + 2g_1x + 2f_1y + c_1 = 0,$$
$$x^2 + y^2 + 2g_2x + 2f_2y + c_2 = 0$$

with coefficients in F_j. Subtracting, $(x_{j+1},\ y_{j+1})$ satisfies the equation

$$2(g_1 - g_2)x + 2(f_1 - f_2)y + (c_1 - c_2) = 0.$$

We cannot have $g_1 = g_2$ and $f_1 = f_2$, for then the circles would be concentric, and would not intersect. Thus this case reduces to the previous one. □

Although the proof of this theorem may appear to be rather lengthy, you should note that almost all the field theory appears in the first paragraph: the rest is coordinate geometry of a particularly simple kind.

Exercise

5.3   If $P = (x, y)$ is constructible, let $P_{\mathbb{C}} = x + iy$, and let $S_{\mathbb{C}} = \{P_{\mathbb{C}} : P \in S\}$. Show that $\mathbb{Q}(S_{\mathbb{C}})$ is a field.

Theorem 5.2 *The circle cannot be squared, the cube cannot be duplicated, and the angle $\pi/3$ cannot be trisected.*

Proof   To square the circle requires the construction of a length $1/\sqrt{\pi}$; but this is not possible, since $1/\sqrt{\pi}$ is transcendental.

Similarly, to duplicate the cube requires the construction of a length $\sqrt[3]{2}$; this is algebraic over $\mathbb{Q}$, but $[\mathbb{Q}(\sqrt[3]{2}) : \mathbb{Q}] = 3$, since $x^3 - 2$ is irreducible over $\mathbb{Q}$.

Let $e(2\pi i/9) = \alpha = a + ib$. If the angle $\pi/3$ could be trisected, then a would be constructible. But $a^2 + b^2 = 1$ and $2(a + ib)^3 = 1 + i\sqrt{34}$, so that if $c = 2a$ then $c^3 - 3c - 1 = 0$. But this is irreducible, and so this is the minimal polynomial for c over $\mathbb{Q}$. Thus $[\mathbb{Q}(a) : \mathbb{Q}] = 3$, and a is not constructible. □

This means that we cannot construct a regular nine-sided polygon with ruler and compasses. However, if $e^{2\pi i/5} = f + ig$ then $2f^2 + 2f - 1 = 0$ so that $f = 1 - \sqrt{3}/2$; thus $e^{2\pi i/5}$ is constructible, and we can construct a regular pentagon with ruler and compasses. What other regular polygons can we construct? We shall answer this question later, in Chapter 13.

6

Splitting Fields

6.1 Introduction

Suppose first that $f \in \mathbb{Q}[x]$. As we have seen, we may factorize f in an essentially unique way into irreducible factors. Further, f can be written as λg, where $\lambda \in \mathbb{Q}$ and g is a primitive element of $\mathbb{Z}[x]$. By Gauss' lemma, f is irreducible in $\mathbb{Q}[x]$ if and only if ig is irreducible in $\mathbb{Z}[x]$. This is as far as factorization can go in $\mathbb{Q}[x]$, or $\mathbb{Z}[x]$.

In contrast, $\mathbb{Q}$ is a subfield of $\mathbb{C}$ and we can consider f as an element of $\mathbb{C}[x]$. Now the field $\mathbb{C}$ has the remarkable property that any non-unit element of $\mathbb{C}[x]$ can be written as a product of linear factors. This is, of course, an immediate consequence of the fact that any non-constant polynomial p in $\mathbb{C}[x]$ has a root in $\mathbb{C}$.[1] This fact (the 'fundamental theorem of algebra') is usually proved by complex function theory: if p had no root, $1/p$ would be a non-constant bounded analytic function on $\mathbb{C}$, contradicting Liouville's theorem. (You may feel that there is too much analysis in this. Some analysis is certainly needed, since the real field $\mathbb{R}$ is an analytic construction. Be patient, and be sure to tackle Exercise 9.11 in due course.)

If we consider f as an element of $\mathbb{C}[x]$, then we can write

$$f = \lambda(x - \alpha_1) \ldots (x - \alpha_n),$$

where λ is a rational number and $\alpha_1, \ldots, \alpha_n$ are complex numbers. Each α_i is algebraic over $\mathbb{Q}$, since $f(\alpha_i) = 0$. Thus, if $L = \mathbb{Q}(\alpha_1, \ldots, \alpha_n)$, L is algebraic over $\mathbb{Q}$ and L is finitely generated over $\mathbb{Q}$, and so $[L : \mathbb{Q}] < \infty$ (Theorem 4.7). Further, f factorizes into linear factors over L. As far as f is concerned, then, L is large enough for our purposes.

---

[1] Three proofs are given in D.J.H. Garling, *A Course in Mathematical Analysis*, Vol. II (Cambridge University Press).

The above argument works because of the special properties of the complex field $\mathbb{C}$. Our aim in this chapter is to show how, starting with an element f of $K[x]$, where K is an arbitrary field, we can construct a finite extension $L : K$ such that f factorizes into linear factors over L.

Exercise

6.1   Suppose that f is an irreducible polynomial in $\mathbb{R}[x]$. Show that degree $f \leqslant 2$.

6.2 Splitting Fields

Suppose that K is a field, that $f \in K[x]$ and that $L : K$ is an extension. We say that f *splits over* L if we can write

$$f = \lambda(x - \alpha_1)\ldots(x - \alpha_n)$$

where $\alpha_1, \ldots, \alpha_n$ are in L and $\lambda \in K$.

We say that $L : K$ is a *splitting field extension for f over K* (or, when it is clear what K is, that L is a *splitting field* for f) if, first, f splits over L and, second, there is no proper subfield L' of L containing K such that f splits over L'. This last condition ensures that the extension $L : K$ is an economical one for f.

If we can find an extension over which f splits, we can find a splitting field:

Theorem 6.1 *Suppose that $L : K$ is an extension and that $f \in K[x]$ splits over L as*

$$f = \lambda(x - \alpha_1)\ldots(x - \alpha_n).$$

Then $K(\alpha_1, \ldots, \alpha_n)$ is a splitting field for f.

Proof f certainly splits over $K(\alpha_1, \ldots, \alpha_n)$. Suppose that $K(\alpha_1, \ldots, \alpha_n) \supseteq K' \supseteq K$ and that f splits over K':

$$f = \lambda'(x - \alpha_1')\ldots(x - \alpha_n').$$

As factorization in $L[x]$ is essentially unique, for each i we have $\alpha_i = \alpha_j'$, for some j, and so $\alpha_i \in K'$. Consequently $K' \supseteq K(\alpha_1, \ldots, \alpha_n)$ and so K' is not a proper subfield of $K(\alpha_1, \ldots, \alpha_n)$.   □

Corollary *If $L : K$ is a splitting field extension for $f \in K[x]$ then $L : K$ is a finite algebraic extension.*

How can we construct splitting fields? The key step is the adjunction of a root of an irreducible polynomial.

Theorem 6.2 *Suppose that $f \in K[x]$ is irreducible of degree n. Then there is a simple algebraic extension $K(\alpha) : K$ such that $[K(\alpha) : K] = n$ and $f(\alpha) = 0$.*

Proof We must construct $K(\alpha)$ intrinsically, starting from K and f. Let $j : K \rightarrow K[x]$ be the natural monomorphism, let $L = K[x]/(f)$, and let $q : K[x] \rightarrow L$ be the quotient map. Since f is irreducible, L is a field (by the corollary to Theorem 2.35). Let $i = qj : i$ is a monomorphism of the field K into the field L:

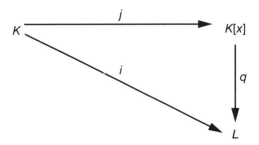

so that (i, K, L) is an extension (remember the original definition). Now let $\alpha = q(x) = x + (f)$. As x generates $K[x]$ over $K, L = K(\alpha)$. Also, since q is a ring homomorphism,

$$f(\alpha) = f(q(x)) = q(f) = 0.$$

Thus α is algebraic over K. Bearing in mind that f is irreducible over K, we see that f must be a scalar multiple of the minimal polynomial m_α of α over K. Thus $[L : K] = n$, by Theorem 4.5.

Note that, although f is irreducible over K, it is not irreducible over $K(\alpha)$: it has a linear factor $x - \alpha$. Factorization is under way, and we can now proceed inductively. $\square$

Theorem 6.3 *Suppose that $f \in K[x]$. Then there exists a splitting field extension $L : K$ for f, with $[L : K] \leqslant n!$.*

Proof We prove this by induction on $n = $ degree f. Of course, if degree $f \leqslant 1$, there is nothing to prove. Suppose that the result holds for any polynomial of degree less than n, over any field K. Suppose that degree $f = n$. We consider two cases.

Case 1. f is not irreducible over K. We can write $f = gh$, where degree $g = s < n$ and degree $h = t < n$. By the inductive hypothesis there is a splitting field $L : K$ for g, with $[L : K] \leqslant s!$. We can write

$$g = \lambda(x - \alpha_1)\ldots(x - \alpha_s)$$

with $\alpha_i \in L$ and $\lambda \in K$. Note that $L = K(\alpha_1, \ldots, \alpha_s)$.

We can now consider h as an element of $L[x]$; by the inductive hypothesis again, there is a splitting field $M : L$ for h, with $[M : L] \leqslant t!$ We can write

$$h = \mu(x - \beta_1)\ldots(x - \beta_t)$$

with $\beta_i \in M$ and $\mu \in L$. Note that $M = L(\beta_1, \ldots, \beta_t) = K(\alpha_1, \ldots, \alpha_s, \beta_1, \ldots, \beta_t)$; as $\lambda\mu$ is the coefficient of x^n in f, $\lambda\mu \in K$. Thus $M : K$ is a splitting field extension for f. Further,

$$[M : K] = [M : L][L : K] \leqslant t! \, s! \leqslant (s + t)! = n! \,.$$

Case 2. f is irreducible over K. Then by Theorem 6.2 there exists a simple algebraic extension $K(\alpha) : K$, with $[K(\alpha) : K] = n$, such that, over $K(\alpha)$,

$$f = (x - \alpha)h$$

where $h \in K(\alpha)[x]$, and degree $h = n - 1$. By the inductive hypothesis, there exists a splitting field extension $L : K(\alpha)$ for h, with $[L : K(\alpha)] \leqslant (n - 1)!$. We can write

$$h = \mu(x - \beta_1)\ldots(x - \beta_{n-1})$$

with $\beta_i \in L$, $\mu \in K(\alpha)$. Note that $L = K(\alpha)(\beta_1, \ldots, \beta_{n-1}) = K(\alpha, \beta_1, \ldots, \beta_{n-1})$. Then

$$f = \mu(x - \alpha)(x - \beta_1)\ldots(x - \beta_{n-1});$$

again, μ is the coefficient of x^n in f, so that $\mu \in K$ and f splits over L. Thus

$$L : K = K(\alpha, \beta_1, \ldots, \beta_{n-1}) : K$$

is a splitting field extension for f. Finally

$$[L : K] = [L : K(\alpha)][K(\alpha) : K] \leqslant (n - 1)! \, n = n! \,. \qquad \square$$

Observe that the proof of Theorem 6.3 is largely a matter of induction; the field theory occurs in Theorem 6.2.

Nevertheless, Theorem 6.3 is a major achievement: we can now produce a splitting field for *any* polynomial over *any* field. Notice that there can be some freedom of action in Theorem 6.3 (in the way we consider factors in the case where f is not irreducible); there may also be other ways to produce splitting fields: can these be essentially different? We shall answer this important question in the next section.

Exercises

6.2   Suppose that $\alpha \in \mathbb{C}$ and that $\alpha^3 + 4\alpha^2 + \alpha + 1 = 0$. Express $(1 - \alpha)^{-2}$ as a linear combination of $1, \alpha$ and α^2.

6.3   Suppose that $f \in K[x]$ is irreducible of degree n and that $L : K$ is a finite extension for which $[l : K]$ and n have no common factor. Can f have a root in L?

6.4   Suppose that $f = x^2 + 2x + 6 \in \mathbb{Z}[x]$. Find a splitting field in $\mathbb{C}[x]$. What if $f \in \mathbb{Z}_7[x]$?

6.5   Suppose that $M : K$ and $M : L$ are field extensions, with $[M : K] < \infty$. Let KL be the field generated by K and L. Show that $[KL : K] \le [L : K \cap L]$.

6.6   Suppose that $[L : K]$ is a splitting field extension for a polynomial $f \in K[x]$ of degree n. Show that L is uniquely determined by $n - 1$ roots of f. Is it uniquely determined by $n - 2$ roots of f?

6.7   Show that $f = x^3 - x + 1$ is irreducible in $\mathbb{Z}_3[x]$. Show that if ζ is a root of f in a splitting field extension, then $\zeta + 1$ and $\zeta - 1$ are also roots. Construct a splitting field extension, and write out its multiplication table.

6.8   Suppose that K is a field over which $x^n - 1$ splits, and suppose that $K(t) : K$ is a simple transcendental extension. Show that $x^n - t$ is irreducible in $K(t)[x]$. Construct a splitting field extension for $x^n - t$ by considering another simple transcendental extension $K(s) : K$ and a monomorphism $i : K(t) \to K(s)$ which fixes K and sends t to s^n.

6.3 The Extension of Monomorphisms

In this section, we shall show that a splitting field extension of a polynomial is essentially unique. In the process, we shall prove some of the most important results of the theory. As the theory develops in the succeeding chapters, it will, I hope, become clear why these results are so important. Two more remarks are in order. First, algebra is not just the study of sets with some algebraic structure, but the study of such sets and of mappings between them which respect the structure: in this section we begin to consider such mappings. Second, although the results are important, the proofs are natural and easy: the relationship between 'difficulty' and 'importance' is a curious one.

   Let us recall that if i is a ring homomorphism from a field K into a field L then i is necessarily a monomorphism, so that i is an isomorphism of K onto $i(K)$. Further if

$$f = a_0 + a_1 x + \cdots + a_n x^n \in K[x]$$

then $i(f) = i(a_0) + i(a_1)x + \cdots + i(a_n)x^n \in i(K)[x] \subseteq L[x]$; thus i extends
to a monomorphism (which we again denote by i) from $K[x]$ into $L[x]$, and i
is an isomorphism of $K[x]$ onto $i(K)[x]$.

We begin by considering simple algebraic extensions.

Theorem 6.4 *Suppose that* $K(\alpha)$: K *is a simple extension and that* α
is algebraic over K, *with minimal polynomial* m_α. *Suppose that* i *is a
monomorphism from* K *into a field* L *and that* $\beta \in L$. *Then a necessary and
sufficient condition for there to be a monomorphism* j *from* $K(\alpha)$ *to* L *with*
$j(\alpha) = \beta$ *and* $j|_K = i$ *is that* $i(m_\alpha)(\beta) = 0$. *If the condition is satisfied then*
j *is unique.*

Proof   Necessity. This is rather trivial. If j exists then

$$i(m_\alpha)(\beta) = j(m_\alpha)(j(\alpha)) = j(m_\alpha(\alpha)) = j(0) = 0.$$

Sufficiency. Suppose that the condition is satisfied. Let $K' = i(K)$. Then i :
$K \to K'$ is an isomorphism, which extends to an isomorphism $i \colon K[x] \to$
$K'[x]$. As $i(m_\alpha)(\beta) = 0$, β is algebraic over K'. We now use the evaluation
maps to construct the following diagram:

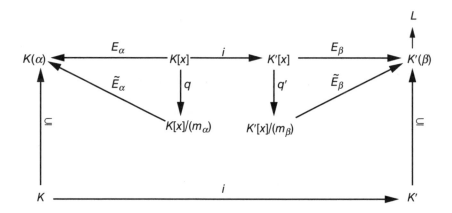

Here E_α and E_β are the evaluation maps, q and q' are quotient maps and $\tilde{E}_\alpha$
and $\tilde{E}_\beta$ are isomorphisms.

Now $i(m_\alpha)$ is monic, and is irreducible over K' (since i is an isomorphism
of $K[x]$ onto $K'[x]$ which sends K to K') and $i(m_\alpha)(\beta) = 0$, by hypothesis.

Consequently $m_\beta = i(m_\alpha)$, and so (m_α) is the kernel of $q'i$. Thus by Theorem 2.1 there exists an isomorphism

$$\tilde{\imath} \colon K[x]/(m_\alpha) \to K'[x]/(m_\beta)$$

such that $q'i = \tilde{\imath}q$. Now let

$$j = \tilde{E}_\beta \tilde{\imath} (\tilde{E}_\alpha)^{-1}.$$

j is an isomorphism of $K(\alpha)$ onto $K'(\beta)$, and so it is a monomorphism of $K(\alpha)$ into L. Also

$$j(\alpha) = \tilde{E}_\beta \tilde{\imath} (\tilde{E}_\alpha)^{-1}(\alpha) = \tilde{E}_\beta \tilde{\imath} q(x)$$
$$= \tilde{E}_\beta q'i(x) = E_\beta(i(x)) = \beta$$

and if $k \in K$

$$j(k) = \tilde{E}_\beta \tilde{\imath} (\tilde{E}_\alpha)^{-1}(k) = \tilde{E}_\beta \tilde{\imath} q(k)$$
$$= \tilde{E}_\beta q'i(k) = E_\beta(i(k)) = i(k),$$

so that j has the properties that we are looking for.

Finally, if j' is another monomorphism of $K(\alpha)$ with the required properties, then the set

$$F = \{\gamma \colon j(\gamma) = j'(\gamma)\}$$

is a subfield of $K(\alpha)$. It contains K and α, and so $F = K(\alpha)$ and j is unique.
$\square$

This theorem can be proved more quickly: it is not really necessary to show that $i(m_\alpha) = m_\beta$. But this proof shows how rigidly j is determined: we have built a strong bridge between $K(\alpha)$ and $K'(\beta)$.

Inspection of the diagram and the proof gives the following corollary:

Corollary 1 *Suppose that $K(\alpha) : K$ and $K'(\alpha') : K'$ are simple extensions, and that α is algebraic over K, α' algebraic over K'. Suppose that $i \colon K \to K'$ is an isomorphism. Then there exists an isomorphism $j \colon K(\alpha) \to K'(\alpha')$ with $j(\alpha) = \alpha'$ and $j|_K = i$ if and only if $i(m_\alpha) = m_{\alpha'}$. If so, j is unique.*

Corollary 2 *Suppose that $K(\alpha) : K$ is simple and that α is algebraic over K. Suppose that $i \colon K \to L$ is a monomorphism, and that $i(m_\alpha)$ has r distinct roots in L. Then there are exactly r distinct monomorphisms $j \colon K(\alpha) \to L$ with $j|_K = i$.*

We now consider splitting fields.

Theorem 6.5 *Suppose that $\Sigma : K$ is a splitting field extension for a polynomial f in $K[x]$ and that i is a monomorphism from K into a field L. Then a necessary and sufficient condition for there to be a monomorphism j from Σ into L with $j|_K = i$ is that $i(f)$ splits over L.*

Proof   Necessity. As f splits over Σ, we can write

$$f = \lambda(x - \alpha_1)\ldots(x - \alpha_n),$$

with $\lambda \in K, \alpha_1, \ldots, \alpha_n \in \Sigma$. Then

$$i(f) = j(f) = i(\lambda)(x - j(\alpha_1))\ldots(x - j(\alpha_n))$$

so that $i(f)$ splits over L.

Sufficiency. Once again we argue by induction on degree $f = n$. The result is true when $n = 1$, for then $\Sigma = K$, and we take $j = i$. Suppose that the result holds for any splitting field extension $\Sigma' : K'$ for any polynomial of degree less than n over any field K', and for any monomorphism i' from K' into L. Suppose that degree $f = n$, and that $i(f)$ splits over L.

As $\Sigma : K$ is a splitting field extension for f over K, we can write

$$f = \lambda(x - \alpha_1)\ldots(x - \alpha_n),$$

with $\alpha_i \in \Sigma$ and $\lambda \in K.\alpha_1$ is algebraic over K; let m be its minimal polynomial over K. Then $f = mg$, and m is irreducible over K. By relabelling $\alpha_1, \ldots, \alpha_n$ if necessary, we can suppose that

$$m = (x - \alpha_1)(x - \alpha_2)\ldots(x - \alpha_r).$$

Now $i(f) = i(m)i(g)$; as $i(f)$ splits over L, $i(m)$ must split over L too. We can write

$$i(m) = (x - \beta_1)\ldots(x - \beta_r).$$

We are now in a position to apply Theorem 6.4: $K(\alpha_1) : K$ is a simple algebraic extension, and α_1 has minimal polynomial m. Also $i(m)(\beta_1) = 0$. There therefore exists a unique monomorphism j_1 from $K(\alpha_1)$ to L such that $j_1(\alpha_1) = \beta_1$ and $j_1|_K = i$:

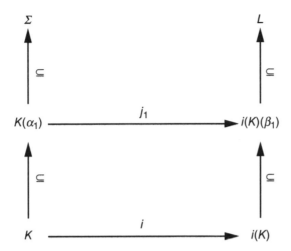

We now consider f as an element of $K(\alpha_1)[x]$. We can write $f = (x-\alpha_1)h$, where $h \in K(\alpha_1)[x]$, and h splits over Σ:

$$h = \lambda(x - \alpha_2)\dots(x - \alpha_n).$$

Also $\Sigma = K(\alpha_1)(\alpha_2, \dots, \alpha_n)$, and so Σ is a splitting field for h over $K(\alpha_1)$. As degree $h = n - 1$, we can apply the inductive hypothesis: there exists a monomorphism j from Σ to L such that $j|_{K(\alpha_1)} = j_1$. This completes the proof. $\qquad\square$

Before we establish some corollaries, let us make three remarks. First, like Theorem 6.4, this is an *extension* theorem: we extend the mapping i. Second, unlike Theorem 6.4, the extension need not be unique: we could map α_1 to any of $\beta_1, \dots, \beta_r$. Third, although the extension need not be unique, there are obviously some limitations on the number of extensions that there can be. This is a topic to which we shall pay much attention later on.

Corollary 1 *Suppose that $i: K \to K'$ is an isomorphism and that $f \in K[x]$. Suppose that $\Sigma : K$ is a splitting field extension for f, $\Sigma' : K'$ a splitting field extension for $i(f)$. Then there exists an isomorphism $j: \Sigma \to \Sigma'$ such that $j|_K = i$.*

Proof If we apply the theorem to the mapping i, considered as a monomorphism from K to Σ', it follows that there exists a monomorphism j from Σ to Σ' which extends i. We can write

$$f = \lambda(x - \alpha_1)\dots(x - \alpha_n),$$

with $\alpha_1, \ldots, \alpha_n$ in Σ and λ in K. Then

$$j(f) = i(\lambda)(x - j(\alpha_1)) \ldots (x - j(\alpha_n)),$$

so that, using Theorem 6.1, it follows that

$$\Sigma' = K'(j(\alpha_1), \ldots, j(\alpha_n)) \subseteq j(\Sigma),$$

and j is onto.   □

This leads to the following fundamentally important result:

Theorem 6.6 *Suppose that $f \in K[x]$ is irreducible and that $\Sigma : K$ is a splitting field extension for f. If α and β are roots of f in Σ, there is an automorphism $\sigma : \Sigma \to \Sigma$ such that $\sigma(\alpha) = \beta$ and σ fixes K.*

Proof   We may suppose that f is monic: then f is the minimal polynomial for α and β over K. By Corollary 1 of Theorem 6.4, there is an isomorphism $\tau : K(\alpha) \to K(\beta)$ with $\tau(\alpha) = \beta$ and $\tau(k) = k$ for $k \in K$. Now $\Sigma : K(\alpha)$ is a splitting field extension for f over $K(\alpha)$, and $\Sigma : K(\beta)$ is a splitting field extension for f over $K(\beta)$. The result now follows from Corollary 1.   □

Exercise

6.9   The complex numbers $i\sqrt{3}$ and $1 + i\sqrt{3}$ are roots of the quartic $f = x^4 - 2x^3 + 7x^2 - 6x + 12$. Does there exist an automorphism σ of the splitting field extension for f over $\mathbb{Q}$ with $\sigma(i\sqrt{3}) = 1 + i\sqrt{3}$?

6.4 Some Examples

We now consider some examples of splitting fields. First let us consider polynomials in $\mathbb{Q}[x]$. If $f \in \mathbb{Q}[x]$ then, as we saw at the beginning of this chapter, f splits over $\mathbb{C}[x]$, and we can, and usually shall, consider the splitting field of f as a subfield of $\mathbb{C}$. Such a field is called a number field. Alternatively, we can make the constructions of Theorem 6.2 and 6.3. Corollary 1 to Theorem 6.5 then says that the splitting field that we obtain is essentially the same.

Example 6.4.1 $f = x^p - 2$ in $\mathbb{Q}[x]$ *(with p a prime).*
   f is irreducible, by Eisenstein's criterion, and there is one real positive root $2^{1/p}$. f is the minimal polynomial of $2^{1/p}$, so that $[\mathbb{Q}(2^{1/p}) : \mathbb{Q}] = p$. If α is any root of f in $\mathbb{C}$, then $(\alpha/2^{1/p})^p = \alpha^p/2 = 1$, so that $\alpha = 2^{1/p}\omega$, where ω is a root of $x^p - 1$. $x^p - 1$ is not irreducible, as

$$x^p - 1 = (x - 1)(x^{p-1} + x^{p-2} + \cdots + 1).$$

*Now $x^{p-1} + x^{p-2} + \cdots + 1$ is irreducible over $\mathbb{Q}$ (Exercise 2.39), so that if ω
is any root of $x^p - 1$ other than 1 then $[\mathbb{Q}(\omega) : \mathbb{Q}] = p - 1$. The map $n \to \omega^n$
is a homomorphism of $\mathbb{Z}$ into the multiplicative group $\mathbb{C}^*$, with kernel $p\mathbb{Z}$, and
so the complex numbers $1, \omega, \ldots, \omega^{p-1}$ must be distinct. They are all roots of
$x^p - 1$, so that*

$$x^p - 1 = (x - 1)(x - \omega) \ldots (x - \omega^{p-1})$$

and $\mathbb{Q}(\omega) : \mathbb{Q}$ is a splitting field extension for $x^p - 1$.

Now our original polynomial f splits over $\mathbb{Q}(\omega, 2^{1/p})$ since it has roots

$$2^{1/p}, \omega 2^{1/p}, \ldots, \omega^{p-1} 2^{1/p}.$$

*Further, any splitting field must contain $2^{1/p}$, and must also contain
$\omega = \omega 2^{1/p}/2^{1/p}$. Thus $\mathbb{Q}(\omega, 2^{1/p}) : \mathbb{Q}$ is the splitting field extension for f.*

*What is $[\mathbb{Q}(\omega, 2^{1/p}) : \mathbb{Q}]$? In order to answer this, consider the following
diagram:*

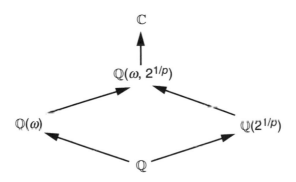

*Here (and later, when we consider similar diagrams) rising arrows represent
inclusion mappings.*

By the tower law,

$$[\mathbb{Q}(2^{1/p}) : \mathbb{Q}] \big| [\mathbb{Q}(\omega, 2^{1/p}) : \mathbb{Q}] \quad \text{and} \quad [\mathbb{Q}(\omega) : \mathbb{Q}] \big| [\mathbb{Q}(\omega, 2^{1/p}) : \mathbb{Q}].$$

*As $[\mathbb{Q}(2^{1/p}) . \mathbb{Q}] = p$ and $[\mathbb{Q}(\omega) : \mathbb{Q}] = p - 1$, this means that $[\mathbb{Q}(\omega, 2^{1/p}) :
\mathbb{Q}] \geqslant p(p - 1)$. However if m is the minimal polynomial of $2^{1/p}$ over $\mathbb{Q}(\omega)$,
m divides $x^p - 2$ in $\mathbb{Q}(\omega)[x]$, and so*

$$\text{degree } m = [\mathbb{Q}(\omega, 2^{1/p}) : \mathbb{Q}(\omega)] \leqslant p.$$

Thus, by the tower law,

$$[\mathbb{Q}(\omega, 2^{1/p}) : \mathbb{Q}] = [\mathbb{Q}(\omega, 2^{1/p}) : \mathbb{Q}(\omega)][\mathbb{Q}(\omega) : \mathbb{Q}]$$
$$\leqslant p(p-1),$$

and so $[\mathbb{Q}(\omega, 2^{1/p}) : \mathbb{Q}] = p(p-1)$.
   This implies that degree $m = p$, and so $x^p - 2$ is irreducible over $\mathbb{Q}(\omega)$.

This example has many important features. It is perhaps a bit more complicated than one might imagine. Notice that the pth roots of unity (the roots of $x^p - 1$) played an important role. Notice also that we picked one of them (other than 1): had we picked another, the result would have been the same. (Can you formalize this, using Theorem 6.6?) Notice also that the argument could have been simplified by appealing to Exercise 4.7.

Example 6.4.2 $f = x^6 - 1$ *in $\mathbb{Q}[x]$.*
   f factorizes as

$$f = (x - 1)(x^2 + x + 1)(x + 1)(x^2 - x + 1).$$

If ω is a root of $x^2 + x + 1$ then

$$f = (x - 1)(x - \omega)(x - \omega^2)(x + 1)(x + \omega)(x + \omega^2).$$

Thus $\mathbb{Q}(\omega) : \mathbb{Q}$ is a splitting field extension for f and $[\mathbb{Q}(\omega) : \mathbb{Q}] = 2$.

Example 6.4.3 $f = x^6 + 1$ *in $\mathbb{Q}[x]$.*
   The roots of f in $\mathbb{C}$ are $i, i\omega, i\omega^2, -i, -i\omega, -i\omega^2$. Thus, arguing as before, $\mathbb{Q}(i, \omega) : \mathbb{Q}$ is the splitting field extension for f, and we have the following diagram:

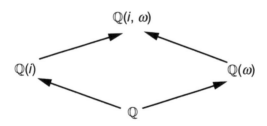

Now we can take $\omega = (-1 + \sqrt{3}\,i)/2$, so that $\omega \notin \mathbb{Q}(i)$ (which consists of all complex numbers of the form $r + is$, with r and s in $\mathbb{Q}$). Thus $\mathbb{Q}(i) \neq \mathbb{Q}(\omega)$ and both $\mathbb{Q}(i)$ and $\mathbb{Q}(\omega)$ are proper subfields of $\mathbb{Q}(i, \omega)$. It is now easy to conclude that $[\mathbb{Q}(i, \omega) : \mathbb{Q}] = 4$.

We now consider examples over more general fields. (To what extent do we use the fact that we are considering polynomials over the rationals in Examples 6.4.1 to 6.4.3?)

Exercises

6.10 Suppose that $M : L$ and $L : K$ are extensions, and that $\alpha \in M$ is algebraic over K. Does $[L(\alpha) : L]$ always divide $[K(\alpha) : K]$?

6.11 Write down all monic cubic polynomials in $\mathbb{Z}_2[x]$, factorize them completely and construct a splitting field for each of them. Which of these fields are isomorphic?

6.12 Find a splitting field extension $K : \mathbb{Q}$ for each of the following polynomials over $\mathbb{Q}$: $x^4 - 5x^2 + 6$, $x^4 + 5x^2 + 6$, $x^4 - 5$. In each case determine the degree $[K : \mathbb{Q}]$ and find α such that $K = \mathbb{Q}(\alpha)$.

6.13 Find a splitting field extension $K : \mathbb{Q}$ for each of the following polynomials over $\mathbb{Q}$: $x^4 + 1$, $x^4 + 4$, $(x^4 + 1)(x^4 + 4)$, $(x^4 - 1)(x^4 + 4)$. In each case determine the degree $[K : \mathbb{Q}]$ and find α such that $K = \mathbb{Q}(\alpha)$.

6.14 Suppose that $L : K$ is a splitting field extension for a polynomial of degree n. Show that $[L : K]$ divides $n!$.

6.15 Find a splitting field extension for $x^3 - 5$ over $\mathbb{Z}_7, \mathbb{Z}_{11}$ and $\mathbb{Z}_{13}$.

7

Normal Extensions

7.1 Basic Properties

There are many field extensions. How do we recognize when an extension $L : K$ is a splitting field extension? For this we need the notion of a normal extension.

An extension $L : K$ is said to be *normal* if it is algebraic and whenever f is an irreducible polynomial in $K[x]$ then *either* f splits over L *or* f has no roots in L. Clearly an algebraic extension $L : K$ is normal if and only if the minimal polynomial over K of each element of L splits over L.

The word 'normal' is one of the most overworked words in mathematical terminology (normal subgroups, normal topological spaces, ...). We shall see in Theorem 11.8 that this is a good use of the word.

In order to characterize normality, we need to extend the definition of a splitting field. Suppose that K is a field, and that S is a subset of $K[x]$. We say that an extension L of K is a *splitting field extension for* S if each f in S splits over L, and if $L \supseteq L' \supseteq K$ and each f in S splits over L', then $L' = L$.

If S is a finite set $\{f_1, \ldots, f_n\}$ then $L : K$ is a splitting field extension for S if and only if it is a splitting field extension for $g = f_1 \ldots f_n$; thus the new definition is only of interest if S is infinite.

Theorem 7.1 *An extension $L : K$ is normal if and only if it is a splitting field extension for some $S \subseteq K[x]$.*

Proof Suppose first that $L : K$ is normal. $L : K$ is algebraic: let $S = \{m_\alpha : \alpha \in L\}$ be the set of minimal polynomials over K of elements of L. By hypothesis, each f in S splits over L, and clearly S splits over no proper subfield of L.

Conversely, suppose that $L : K$ is a splitting field extension or S. Let A denote the set of roots in L of polynomials in S. Then clearly $L = K(A)$, and so $L : K$ is algebraic, by Corollary 2 to Theorem 4.7.

Suppose that $\beta \in L$ and that m is its minimal polynomial over K. We must show that m splits over L. First we reduce the problem to one concerning finite extensions. As $\beta \in K(A)$, there exist $\alpha_1, \ldots, \alpha_n$ in A such that $\beta \in K(\alpha_1, \ldots, \alpha_n)$. There exist $f_1, \ldots, f_n$ in S such that α_i. is a root of f_i, for $1 \leq i \leq n$. Each f_i splits over L. Let R be the set of roots of $g = f_1 \ldots f_n$. Then $K(R) : K$ is a splitting field extension for g and $\beta \in K(R)$. We now consider m as an element of $K(R)[x]$ and construct a splitting field extension $H : K(R)$ for m. Let γ be another root of m in H. We must show that in fact $\gamma \in K(R)$.

We have the following diagram, where upward-pointing arrows denote inclusions:

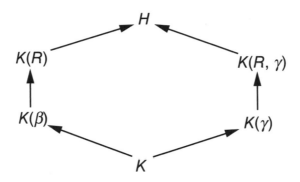

m is the minimal polynomial of both β and γ over K, so that $[K(\beta) : K] = [K(\gamma) : K] = $ degree m. Also, by the corollary to Theorem 6.4, there is an isomorphism τ of $K(\beta)$ onto $K(\gamma)$ which sends β to γ and which fixes K. As τ fixes K, $\tau(g) = g$.

Now $K(R) : K(\beta)$ is a splitting field extension for g over $K(\beta)$, and $K(R, \gamma) : K(y)$ is a splitting field extension for $\tau(g) = g$ over $K(\gamma)$, so that by Corollary 1 to Theorem 6.5 there is an isomorphism σ of $K(R)$ onto $K(R, \gamma)$ such that $\sigma|_{K(\beta)} = \tau$. This means that $[K(R) : K(\beta)] = [K(R, \gamma) : K(\gamma)]$, and so by the tower law

$$[K(R) : K] = [K(R) : K(\beta)][K(\beta) : K]$$
$$= [K(R, \gamma) : K(\gamma)][K(\gamma) : K]$$
$$= [K(R, \gamma) : K].$$

But $K(R) \subseteq K(R, \gamma)$, and so we must have that $K(R) = K(R, \gamma)$. Consequently, $\gamma \in K(R)$.   □

The case of finite extensions is particularly important:

Corollary 1 *A finite extension* $L : K$ *is normal if and only if* $L : K$ *is a splitting field extension for some* $g \in K[x]$.

For if $L : K$ is normal and finite, and $\alpha_1, \ldots, \alpha_n$ is a basis for L over K, then $L : K$ is a splitting field extension for $g = m_{\alpha_1} m_{\alpha_2} \ldots m_{\alpha_n}$.

Let $L = K(\alpha_1, \ldots, \alpha_n)$, let m_i be the minimal polynomial of α_i over K, and let $g = m_1 \ldots m_n$. Then $L : K$ is normal if and only if $L : K$ is a splitting field extension for g.

Suppose that $L : K$ is algebraic. An extension $F : L$ is a *normal closure* for $L : K$ if $F : K$ is normal, and if $F : M : L$ is a tower and $M : K$ is normal, then $M = F$.

Corollary 2 *If* $L : K$ *is finite, it has a finite normal closure* $F : L$.

With the same notation as in Corollary 1, let $F : L$ be a splitting field extension for g over L. Then $F : K$ is a splitting field extension for g over K so that $F : K$ is normal. If $F : M : L$ is a tower and $M : K$ is normal, then each m_α splits over M, and so g splits over M; therefore $M = F$.

Corollary 3 *If* $L : K$ *is normal and* M *is an intermediate field then* $L : M$ *is normal.*

For there exists $S \subseteq K[x]$ such that $L : K$ is a splitting field extension for S. If we consider S as a subset of $M[x]$, $L : M$ is a splitting field extension for S.

Exercises

7.1 Suppose that $L : K$ and that M and N are normal extensions of K contained in L. Show that $M \cap N$ and MN (the field generated by M and N) are both normal extensions of K.

7.2 Show that every algebraic extension has a normal closure.

7.3 Suppose that $L : K$ is algebraic. Show that there is a greatest intermediate field M for which $M : K$ is normal.

7.4 Suppose that $L : K$ and that M_1 and M_2 are intermediate fields. Show that if $M_1 : K$ and $M_2 : K$ are normal then so are $K(M_1, M_2) : K$ and $M_1 \cap M_2 : K$.

7.2 Monomorphisms and Automorphisms

We have just seen that if $L : K$ is normal and M is an intermediate field then $L : M$ is normal. However, there is no reason why $M : K$ should be normal. For example, if ω is a complex cube root of 1 then $\mathbb{Q}(2^{1/3}, \omega) : \mathbb{Q}$ is normal, since it is the splitting field for $f = x^3 - 2$, while $\mathbb{Q}(2^{1/3}) : \mathbb{Q}$ is not, since f is irreducible and has one root in $\mathbb{Q}(2^{1/3})$ but does not split over $\mathbb{Q}(2^{1/3})$.

It is important to be able to recognize when $M : K$ is normal. In the next theorem we give necessary and sufficient conditions for this, for finite extensions.

Theorem 7.2 *Suppose that $L : K$ is a finite normal extension and that M is an intermediate field. The following are equivalent:*

(i) *$M : K$ is normal;*
(ii) *if σ is an automorphism of L which fixes K then $\sigma(M) \subseteq M$;*
(iii) *if σ is an automorphism of L which fixes K then $\sigma(M) = M$.*

Proof Suppose first that $M : K$ is normal, and that σ is an automorphism of L which fixes K. Suppose that $\alpha \in M$ and let m be the minimal polynomial for α over K. Then $m(\sigma(\alpha)) = \sigma(m(\alpha)) = 0$, so that $\sigma(\alpha)$ is a root of m. As m splits over M, $\sigma(\alpha) \in M$, and so $\sigma(M) \subseteq M$. Thus (i) implies (ii). □

Since $[\sigma(M) : K] = [M : K]$ it is clear that (ii) implies (iii).

Suppose now that (iii) holds. As $L : K$ is normal, $L : K$ is the splitting field extension for some $g \in K[x]$, by Corollary 1 to Theorem 7.1. Suppose that $\alpha \in M$. Let m be the minimal polynomial for α over K. As $L : K$ is normal, m splits over L. We must show that m splits over M: that is, that all the roots of m are in M. Let β be any root of m in L. By Theorem 6.4, there exists a monomorphism j from $K(\alpha)$ to $K(\beta)$, fixing K, such that $j(\alpha) = \beta$. Since j fixes K, $j(g) = g$. Now $L : K(\alpha)$ and $L : K(\beta)$ are splitting field extensions for $j(g) = g$. By Corollary 1 to Theorem 6.5, there is an isomorphism $\sigma : L \to L$ which extends j. As σ fixes K, $\sigma(M) = M$. In particular, this means that $\beta = \sigma(\alpha) \in M$.

Exercises

7.5 Suppose that $N : L$ and $N' : L$ are two normal closures of $L : K$. Show that there is an isomorphism j of N onto N' such that $j(l) = l$ for $l \in L$.

7.6 Suppose that $L : K$ is a finite normal extension and that f is an irreducible polynomial in $K[x]$. Suppose that g and h are irreducible monic factors of f in $L[x]$. Show that there is an automorphism σ of L which fixes K such that $\sigma(g) = h$.

7.7   Suppose that $L : K$ is algebraic. Show that the following are equivalent:

   (i) $L : K$ is normal;

   (ii) if j is any monomorphism from L to $\overline{L}$ which fixes K then $j(L) \subseteq L$;

   (iii) if j is any monomorphism from L to $\overline{L}$ which fixes K then $j(L) = L$.

8

Separability

8.1 Basic Ideas

Normality is a property that an extension may or may not have. Separability is different; most extensions of interest are separable, and we shall have to work hard to find examples of non-separable extensions. But separability is an important property; it leads to some very important results in Theorems 8.3 and 8.4.

Separability involves several definitions. Suppose first that f is an irreducible polynomial of degree n in $K[x]$ and that $L : K$ is a splitting field extension for f. We say that f is *separable* (over K) if f has n distinct roots in L. Suppose next that f is an arbitrary polynomial in $K[x]$. We say that f is *separable* (over K) if each of its irreducible factors is separable.

Suppose that $L : K$ is an algebraic extension and that $\alpha \in L$. We say that α is *separable* (over K) if its minimal polynomial over K is separable, and say that $L : K$ is *separable* if each α in L is separable over K.

Theorem 8.1 *Suppose that $L : K$ is separable and that M is an intermediate field. Then $L : M$ and $M : K$ are separable.*

Proof   It is obvious that $M : K$ is separable. $\square$

Suppose that $\alpha \in L$. Let m_1 be its minimal polynomial over M, m_2 its minimal polynomial over K. Let $N : M$ be a splitting field extension for m_2, considered as an element of $M[x]$. Since m_2 is separable over K, we can write

$$m_2 = (x - \alpha_1) \ldots (x - \alpha_r)$$

where $\alpha_1, \ldots, \alpha_r$ are distinct elements of N. But $m_1 | m_2$ in $M[x]$, and so in $N[x]$

$$m_1 = (x - \alpha_{i_1}) \ldots (x - \alpha_{i_2})$$

for some $1 \leq i_1 < \cdots < i_s \leq r$. Thus m_1 is separable.

8.2 Monomorphisms and Automorphisms

We have already seen that counting dimension leads to some remarkably strong results. We shall find that counting monomorphisms and automorphisms is equally useful. With this in mind, the results in this section suggest why separability is important.

First we consider simple extensions.

Theorem 8.2 *Suppose that $K(\alpha) : K$ is a simple algebraic extension of degree d. Suppose that $j : K \to L$ is a monomorphism. If α is separable over K and if $j(m_\alpha)$ splits over L then there are exactly d monomorphisms from $K(\alpha)$ to L extending j; otherwise there are fewer than d such monomorphisms.*

Proof By Corollary 2 to Theorem 6.4, there are r such extensions, where r is the number of distinct roots of $j(m_\alpha)$ in L. Now $d = $ degree $m_\alpha = $ degree $j(m_\alpha)$ (Theorem 4.5), so that $r \le d$, and $r = d$ if and only if $j(m_\alpha)$ splits into d distinct linear factors: that is, if and only if $j(m_\alpha)$ is separable over $j(K)$ and $j(m_\alpha)$ splits over L. Clearly α is separable over K if and only if $j(m_\alpha)$ is separable over $j(K)$, and so the result is proved.   □

We now consider the general case.

Theorem 8.3 *Suppose that $K' : K$ is a finite extension of degree d, and that $j : K \to L$ is a monomorphism. If $K' : K$ is separable and $j(m_\alpha)$ splits over L for each α in K', then there are exactly d monomorphisms from K' to L extending j; otherwise, there are fewer than d such monomorphisms.*

Proof We prove this by induction on d. It is trivially true when $d = 1$. Suppose that it is true for all extensions of degree less than d, and that $[K' : K] = d$.   □

Suppose first that the conditions are satisfied. Let $\alpha \in K' \backslash K$. By Theorem 8.2 there are exactly $[K(\alpha) : K]$ monomorphisms from $K(\alpha)$ to L extending j. Let k be one of these. We apply the inductive hypothesis to $K' : K(\alpha)$. First, $[K' : K(\alpha)] < d$. Second, $K' : K(\alpha)$ is separable, by Theorem 8.1. If $\beta \in K'$, let m_β be the minimal polynomial for β over K and let n_β be the minimal polynomial for β over $K(\alpha)$. Then n_β divides m_β in $K(\alpha)[x]$ and so $k(n_\beta)$ divides $k(m_\beta)$ in $L[x]$. But $k(m_\beta)$ splits over $L[x]$, and so $k(n_\beta)$ splits over $L[x]$. Thus the conditions are satisfied, and so k can be extended in $[K' : K(\alpha)]$ ways. It therefore follows from the tower law that j can be extended in d ways.

Suppose next that the conditions are not satisfied. Then there exists α in K' such that $j(m_\alpha)$ has fewer than $[K(\alpha) : K]$ distinct roots in L, and so j can be extended in fewer than $[K(\alpha) : K]$ ways to a monomorphism from $K(\alpha)$ to L, by Corollary 2 to Theorem 6.4. Each of these extensions can be extended to a monomorphism from K' to L in at most $[K' : K(\alpha)]$ ways, by the inductive hypothesis, and so there are fewer than d extensions.

Corollary 1 *Suppose that $L : K$ is finite and that $L = K(\alpha_1, \ldots, \alpha_r)$. If α_i is separable over $K(\alpha_1, \ldots, \alpha_{i-1})$ for $1 \leq i \leq r$, then $L : K$ is separable.*

Proof Let $F : L$ be a normal closure for $L : K$. Let $K_0 = K$, and let $K_j = K(\alpha_1, \ldots, \alpha_j) = K_{j-1}(\alpha_j)$ for $1 \leq j \leq r$. We assert that there are $[K_j : K]$ monomorphisms from K_j into F which fix K. The result is trivially true for $j = 0$. Assume that it is true for $j - 1$, and that i is a monomorphism from K_{j-1} to F which fixes K. Let n_j be the minimal polynomial for α_j over K_{j-1}, and let m_j be the minimal polynomial for α_j over K. Then $n_j | m_j$ in $K_{j-1}[x]$, and so $i(n_j)|i(m_j)$ in $i(K_{j-1})[x]$. But $i(m_j) = m_j$, and m_j splits in $F[x]$, so that $i(n_j)$ splits in $F[x]$. As α_j is separable over K_{j-1}, i can be extended in $[K_j : K_{j-1}]$ ways to a monomorphism from K_j to F, by Theorem 8.2. The assertion therefore follows inductively, using the tower law. But it now follows from Theorem 8.3 that $K_j : K$ is separable, and so, in particular, $L : K$ is separable.   □

Corollary 2 *Suppose that $L : K$ is finite and that $L = K(\alpha_1, \ldots, \alpha_r)$. If each α_i is separable over K then $L : K$ is separable.*

This follows from Corollary 1 and Theorem 8.1.

Corollary 3 *Suppose that $f \in K[x]$ is separable over K and that $L : K$ is a splitting field extension for f. Then $L : K$ is separable.*

Apply Corollary 2 to the roots of f in L.

Corollary 4 *Suppose that $L : K$ is finite, and that $L : M : K$ is a tower. If $L : M$ and $M : K$ are separable, then so is $L : K$.*

Write $M = K(\alpha_1, \ldots, \alpha_r)$, $L = M(\alpha_{r+1}, \ldots, \alpha_s)$, and use Corollary 1 and Theorem 8.1.

Exercise

8.1 Suppose that $L : K$ is finite and that $L' : L$ is a normal closure for $L : K$. Show that $L : K$ is separable if and only if there are exactly $[L : K]$ monomorphisms of L into L' which fix K.

8.3 Galois Extensions

A separable splitting field extension is called a Galois extension.

Theorem 8.4 *Suppose that* $[L : K]$ *is finite. The following are equivalent:*

(i) $L : K$ *is a Galois extension;*
(ii) $l : K$ *is normal and separable;*
(iii) there are $[L : K]$ *automorphisms of* L *which fix* K.

Otherwise there are fewer than $[L : K]$ *automophisms of* L *which fix* K.

Proof This follows from the corollaries to Theorem 7.1 and Theorem 8.3. We shall extend this result in Theorem 9.4. □

8.4 Differentiation

Suppose that f is a non-zero element of $K[x]$ and that $L : K$ is a splitting field extension for f. We say that f has a *repeated root* in L if there exists $\alpha \in L$ and $k > 1$ such that $(x - \alpha)^k | f$ in $L[x]$. The largest possible value of k is the *multiplicity* of the root α.

An irreducible polynomial in $K[x]$ is not separable if and only if it has a repeated root in a splitting field. It is therefore important to be able to recognize when a polynomial has a repeated root.

Suppose that f is a non-zero polynomial in $\mathbb{C}[x]$, and that α is a root of f. How do we tell if α is a repeated root? We differentiate: α is a repeated root if and only if $f'(\alpha) = 0$. Although differentiation has its roots in analysis, the differential operator has strong algebraic properties – in particular, $(fg)' = f'g + fg'$ – and we can define the derivative of a polynomial in a purely algebraic way.

Suppose that

$$f = a_0 + a_1 x + \cdots + a_n x^n \in K[x].$$

We define the derivative

$$Df = a_1 + 2a_2 x + \cdots + n a_n x^{n-1}.$$

Here, as usual, $j a_j = a_j + \cdots + a_j$ (j times).

D is a mapping from $K[x]$ to $K[x]$. As

$$D(f + g) = Df + Dg, \qquad D(\alpha f) = \alpha(Df),$$

D is a K-linear mapping. Also

$$D(x^m x^n) = (m+n)^{m+n-1} = mx^{m-1}x^n + nx^m x^{n-1} = (Dx^m)x^n + x^m(Dx^n),$$

and so, by linearity,

$$D(fg) = (Df)g + f(Dg).$$

Notice also that, if K has non-zero characteristic p, then

$$Dx^p = px^{p-1} = 0.$$

Differentiation provides a test for repeated roots, just as in the case of $\mathbb{C}[x]$.

Theorem 8.5 *Suppose that is a non-zero element of $K[x]$ and that $L : K$ is a splitting field for f. The following are equivalent:*

(i) *f has a repeated root in L;*
(ii) *there exists α in L for which $f(\alpha) = (Df)(\alpha) = 0$;*
(iii) *there exists m in $K[x]$, with degree $m \geq 1$, such that $m|f$ and $m|Df$.*

Proof Suppose that f has a repeated root α in L. Then $f = (x-\alpha)^k g$, where $k > 1$ and $g \in L[x]$. Thus

$$Df = k(x-\alpha)^{k-1}g + (x-\alpha)^k Dg,$$

and so $f(\alpha) = Df(\alpha) = 0$. Thus (i) implies (ii).

Suppose that (ii) holds. Let m be the minimal polynomial of α over K. Then $m|f$ and $m|Df$, and so (iii) holds.

Suppose that (iii) holds. We can write $f = mh$, with h in $K[x]$. As f splits over L, so does m. Let α be a root of m in L. We can write $f = (x-\alpha)q$, with q in $L[x]$. Then

$$Df = q + (x-\alpha)Dq.$$

But $(x-\alpha)|Df$ in $L[x]$, since $m|Df$, and so $(x-\alpha)|q$. Thus $(x-\alpha)^2|f$, and f has a repeated root in L. $\square$

This theorem enables us to characterize irreducible polynomials which are not separable.

Theorem 8.6 *Suppose that $f \in K[x]$ is irreducible. Then f is not separable if and only if* char $K = p > 0$ *and f has the form*

$$f = a_0 + a_1 x^p + a_2 x^{2p} + \cdots + a_n x^{np}.$$

Proof If f is not separable, there exists m in $K[x]$, with degree $m \geq 1$, such that $m \mid f$ and $m \mid Df$. As f is irreducible, f and m are associates. Thus $f \mid Df$; as degree $Df <$ degree f, it follows that $Df = 0$. This can only happen if char $K \neq 0$ and f has the form given in the theorem.

Conversely, if the conditions are satisfied, $Df = 0$ and we can take $f = m$ in Theorem 8.5(iii).   □

Corollary 1 *If* char $K = 0$, *all polynomials in* $K[x]$ *are separable.*

Corollary 2 *Suppose that* char $K = p > 0$ *and that* K *is perfect. Then* K *is separable.*

Proof For if $f = a_0 + a_1 x^p + \cdots + a_n x_n^p$ and $a_j = b_j^p$ for $0 \leq j \leq p$ then $f = (b_0 + b_1 x + \cdots + b_n x^n)^p$, so that f is not irreducible.   □

Exercises

8.2 Suppose that f is a polynomial in $K[x]$ of degree n and that either char $K = 0$ or char $K > n$. Suppose that $\alpha \in K$. Establish *Taylor's formula*:

$$f = f(\alpha) + Df(\alpha)(x - \alpha)$$

$$+ \frac{D^2 f(\alpha)}{2!}(x - \alpha)^2 + \cdots + \frac{D^n f(\alpha)}{n!}(x - \alpha)^n.$$

8.3 Suppose that f is a polynomial in $K[x]$ of degree n and that either char $K = 0$ or char $K > n$. Show that α is a root of multiplicity $r(\leq n)$ if and only if

$$f(\alpha) = Df(\alpha) = \cdots = D^{r-1} f(\alpha) = 0 \quad \text{and} \quad D^r f(\alpha) \neq 0.$$

8.4 Suppose that p is a prime number. By factorizing $x^{p-1} - 1$ over $\mathbb{Z}_p$, show that $(p - 1)! + 1 = 0 \pmod{p}$ (Wilson's theorem).

8.5 Suppose that p is a prime number of the form $4n + 1$. Show that there exists k such that $k^2 + 1 = 0 \pmod{p}$. Show that p is not a prime in $\mathbb{Z} + i\mathbb{Z}$ and show that there exist u and v in $\mathbb{Z}$ such that $u^2 + v^2 = p$.

8.6 Suppose that p is a prime number of the form $4n + 3$. Show that p is a prime in $\mathbb{Z} + i\mathbb{Z}$.

8.5 Inseparable Polynomials

Suppose that

$$f = a_0 + a_1 x^p + \cdots + a_n x^{np}$$

is in $K[x]$. We shall write $f(x) = g(x^p)$, where

$$g = a_0 + a_1x + \cdots + a_nx^n.$$

This is a slight abuse of terminology, which does not lead to any difficulties.

Theorem 8.7 *Suppose that* char $K = p > 0$ *and that*

$$f(x) = g(x^p) = a_0 + a_1x^p + \cdots + x^{np}$$

is monic; then f is irreducible in $K[x]$ if and only if g is irreducible in $K[x]$, and not all of the coefficients a_i are pth powers of elements of K.

Proof If g factorizes as $g = g_1g_2$, then f factorizes as $f(x) = g_1(x^p)g_2(x^p)$: thus if f is irreducible, so is g.

Suppose next that each a_i is a pth power of an element of K: that is, $a_i = b_i^p$, for b_i in K. Then, as before,

$$
\begin{aligned}
f &= b_0^p + b_1^px^p + \cdots + b_n^px^{np} \\
  &= (b_0 + b_1x + \cdots + b_nx^n)^p
\end{aligned}
$$

and so f factorizes. Thus if f is irreducible, not all the a_i can be pth powers of elements of K.

Conversely, suppose that f factorizes. We must show that either g factorizes or that all the a_i are pth powers of elements of K. We can write f as a product of i irreducible factors:

$$f = f_1^{n_1} \cdots f_r^{n_r}$$

where the f_i are monic and irreducible in $K[x]$, f_i and f_j are relatively prime, for $i \neq j$, and $n_1 + \cdots + n_r > 1$. We have to consider two cases.

First suppose that $r > 1$. Then we can write $f = h_1h_2$, with h_1 and h_2 relatively prime (take $h_1 = f_1^{n_1}$).

There exist λ_1 and λ_2 in $K[x]$ such that

$$\lambda_1h_1 + \lambda_2h_2 = 1.$$

Further,

$$0 = Df = (Dh_1)h_2 + h_1(Dh_2).$$

Eliminating h_2, we find that

$$Dh_1 = \lambda_1h_1(Dh_1) - \lambda_2h_1(Dh_2)$$

and so $h_1|Dh_1$. As degree $Dh_1 <$ degree h_1, we must have $Dh_1 = 0$. Similarly $Dh_2 = 0$. Thus we can write

$$h_1(x) = c_0 + c_1 x^p + \cdots + c_s x^{sp} = j_1(x^p),$$
$$h_2(x) = d_0 + d_1 x^p + \cdots + d_t x^{tp} = j_2(x^p)$$

and g factorizes as $g = j_1 j_2$.

Second, suppose that $r = 1$. Then $f = f_1^n$, where f_1 is irreducible, and $n > 1$. Again there are two cases to consider. If $p|n$, we can write $f = h^p$. If

$$h = c_0 + c_1 x + \cdots + c_s x^s$$

then

$$f = h^p = c_0^p + c_1^p x^p + \cdots + c_s^p x^{sp}$$

so that all the a_i are pth powers. If p does not divide n,

$$0 = Df = n(Df_1)f_1^{n-1}$$

and so $Df_1 = 0$. Thus we can write

$$f_1(x) = d_0 + d_1 x^p + \cdots + d_l x^{lp} = g_1(x^p)$$

and $g = (g_1)^n$. $\qquad\qquad\qquad\qquad\qquad\qquad\qquad\qquad\qquad\quad\square$

Bearing in mind the corollary to Theorem 8.6, this means that if we are to find an inseparable polynomial we must consider fields K of non-zero characteristic which are not algebraic over their prime subfields.

With this information, the search is rather short. Let $K = \mathbb{Z}_p(\alpha)$ be the field of rational expressions in α over $\mathbb{Z}_p$. Suppose if possible that $-\alpha = \beta^p$, for some β in K. Then we can write $\beta = f(\alpha)/g(\alpha)$, with f and g in $\mathbb{Z}_p[x]$. Thus

$$-\alpha(g(\alpha))^p = (f(\alpha))^p$$

and so, since α is transcendental,

$$-xg^p = f^p.$$

But $p|$degree(f^p) and p does not divide degree $(-xg^p)$. Thus $-\alpha$ is not a pth power in K, and so $x^p - \alpha$ is irreducible in $K[x]$, by Theorem 8.7. Let $L : K$ be a splitting field extension for $x^p - \alpha$, and let γ be a root of $x^p - \alpha$ in L. Then

$$(x - \gamma)^p = x^p - \gamma^p = x^p - \alpha$$

so that $x^p - \alpha$ fails to be separable in the most spectacular way.

Exercises

8.7   Show that a field K is perfect if and only if every finite extension of K is separable.

8.8   Suppose that char $K = p > 0$ and that f is irreducible in $K[x]$. Show that f can be written in the form $f(x) = g(x^{p^n})$, where n is a non-negative integer and g is irreducible and separable.

8.9   Suppose that char $K = p > 0$ and that $L : K$ is a *totally inseparable* algebraic extension: that is, every element of $L \backslash K$ is inseparable. Show that if $\beta \in L$ then its minimal polynomial over K is of the form $x^{p^n} - \alpha$, where $\alpha \in K$.

8.10 Suppose that char $K = p \neq 0$, that f is irreducible in $K[x]$ and that $L : K$ is a splitting field extension for f. Show that there exists a non-negative integer n such that every root of f in L has multiplicity p^n. (*Hint*: Use Exercise 8.5.)

9

The Fundamental Theorem of Galois Theory

9.1 Field Automorphisms, Fixed Fields and Galois Groups

Recall that a ring homomorphism from a field L_1 to a field L_2 is a monomorphism, and that if $L : K$ is an algebraic field extension and that $\tau : L \to L$ is a homomorphism which is fixed on K, then τ is an automorphism of L (Theorem 4.9).

Galois theory is largely concerned with properties of groups of automorphisms of a field. If L is a field, we denote by Aut L the set of all automorphisms of L. Aut L is a group under the usual law of composition.

Suppose that A is a subset of Aut L. We set

$$\phi(A) = \{k \in L : \sigma(k) = k \text{ for each } \sigma \text{ in } A\}.$$

It is easy to verify that $\phi(A)$ is a subfield of L, which we call the *fixed field* of A. In this way, starting from A we obtain an extension $L : \phi(A)$.

Conversely suppose that $L : K$ is an extension. We denote by $\Gamma(L : K)$ the set of those automorphisms of L which fix K:

$$\Gamma(L : K) = \{\sigma \in \text{Aut } L : \sigma(k) = k \text{ for all } k \text{ in } K\}.$$

When there is no doubt what the larger field L is, we shall write $\gamma(K)$ for $\Gamma(L : K)$. It is again easy to verify that $\Gamma(L : K)$ is a subgroup of Aut L; we call $\Gamma(L : K)$ the *Galois group* of the extension $L : K$. In this case, then, starting from an extension we obtain a set of automorphisms.

In this chapter we shall study this reciprocal relationship in detail.

The operations $A \to \phi(A)$ and $L : K \to \gamma(K)$ establish a *polarity* between sets of automorphisms of L and extensions $L : K$. The next theorem is a standard result for such polarities.

Theorem 9.1 *Suppose that $L : K$ is an extension, and that A is a subset of* Aut L.

(i) $\gamma\phi(A) \supseteq A$;

(ii) $\phi\gamma(K) \supseteq K$;

(iii) $\phi\gamma\phi(A) = \phi(A)$;

(iv) $\gamma\phi\gamma(K) = \gamma(K)$.

Proof If $\sigma \in A$, $\sigma(k) = k$ for each k in $\phi(A)$, so that $\sigma \in \gamma\phi(A)$: this establishes (i). If $k \in K$, $\sigma(k) = k$ for each σ in $\gamma(K)$, so that $k \in \phi\gamma(K)$: this establishes (ii).

If $A_1 \subseteq A_2$, then clearly $\phi(A_1) \supseteq \phi(A_2)$. Thus it follows from (i) that

$$\phi\gamma\phi(A) \subseteq \phi(A);$$

but applying (ii), with $\phi(A)$ in place of K,

$$\phi\gamma\phi(A) \supseteq \phi(A).$$

This establishes (iii).

Similarly if $K_1 \subseteq K_2$, $\gamma(K_1) \supseteq \gamma(K_2)$. Applying this to (ii):

$$\gamma\phi\gamma(K) \subseteq \gamma(K);$$

but applying (i), with $\gamma(K)$ in place of A,

$$\gamma\phi\gamma(K) \supseteq \gamma(K).$$

This establishes (iv). □

Corollary *If A is a subset of* Aut L, *and $\langle A \rangle$ is the subgroup of* Aut L *generated by A, then $\phi(A) = \phi(\langle A \rangle)$.*

For $A \subseteq \langle A \rangle \subseteq \gamma\phi(A)$, by (i), and so

$$\phi(A) \supseteq \phi(\langle A \rangle) \supseteq \phi\gamma\phi(A) = \phi(A), \text{by (iii)}.$$

Because of this, we shall usually restrict attention to *subgroups* of Aut L.

9.2 Linear Independence

If $\tau \in \text{Aut}(L)$, then the restriction of τ to the multiplicative group $(L^*, \times)$ is an L-valued character on L^*. Thus if $\tau_1, \ldots, \tau_n$ are distinct automorphisms of a field L then they are linearly independent L-valued functions on L^*. But we need more.

Suppose now that G is a subgroup of Aut L. If $\lambda \in L$, we define the *trajectory* of λ, $T_G(\lambda)$, to be the element of L^G defined by $T_G(\lambda)(\sigma) = \sigma(\lambda)$.

L^G is a vector space over L; we can also consider it as a vector space over any subfield of L, and in particular as a vector space over $\phi(G)$.

The next theorem is particularly important: it takes a rather curious form, as it is concerned with linear independence over two different fields.

Theorem 9.2 *Suppose that G is a subgroup of* Aut L, *that K is the fixed field of G and that B is a subset of L. Then the following are equivalent:*

(i) *B is linearly independent over K;*
(ii) *$\{T_G(\beta): \beta \in B\}$ is linearly independent over K;*
(iii) *$\{T_G(\beta): \beta \in B\}$ is linearly independent over L.*

Proof Clearly (iii) implies (ii). Suppose that B is not linearly independent over K: there exist distinct $\beta_1, \ldots, \beta_n$ in B, and $k_1, \ldots, k_n$ in K, not all zero, such that

$$k_1\beta_1 + \cdots + k_n\beta_n = 0.$$

Then if $\sigma \in G$,

$$k_1\sigma(\beta_1) + \cdots + k_n\sigma(\beta_n) = \sigma(k_1\beta_1 + \cdots + k_n\beta_n) = 0,$$

so that $k_1T_G(\beta_1) + \cdots + k_nT_G(\beta_n) = 0$, and the set $\{T_G(\beta): \beta \in B\}$ is not linearly independent over K in L^G. Thus (ii) implies (i).

Finally, suppose that the set of trajectories $\{T_G(\beta): \beta \in B\}$ is not linearly independent over L in L^G. There exist $\beta_1, \ldots, \beta_r$ in B, and non-zero $\lambda_1, \ldots, \lambda_r$ in L such that

$$\lambda_1 T_G(\beta_1) + \cdots + \lambda_r T_G(\beta_r) = 0;$$

further we can find $\beta_1, \ldots, \beta_r$ and $\lambda_1, \ldots, \lambda_r$ so that r is as small as possible. In detail, this says that

$$\lambda_1\sigma(\beta_1) + \cdots + \lambda_r\sigma(\beta_r) = 0 \text{ for each } \sigma \text{ in } G. \tag{*}$$

Now if $\tau \in G$ and $\sigma \in G$ then $\tau^{-1}\sigma \in G$, so that

$$\lambda_1\tau^{-1}\sigma(\beta_1) + \cdots + \lambda_r\tau^{-1}\sigma(\beta_r) = 0 \text{ for each } \sigma \text{ in } G.$$

Operate on this equation by τ:

$$\tau(\lambda_1)\sigma(\beta_1) + \cdots + \tau(\lambda_r)\sigma(\beta_r) = 0 \text{ for each } \sigma \text{ in } G. \tag{**}$$

Now multiply (*) by $\tau(\lambda_r)$, (**) by λ_r, and subtract:

$$(\tau(\lambda_r)\lambda_1 - \tau(\lambda_1)\lambda_r)\sigma(\beta_1) + \cdots + (\tau(\lambda_r)\lambda_{r-1} - \tau(\lambda_{r-1})\lambda_r)\sigma(\beta_{r-1}) = 0$$

for each σ in G. Thus

$$(\tau(\lambda_r)\lambda_1 - \tau(\lambda_1)\lambda_r)T_G(\beta_1) + \cdots + (\tau(\lambda_r)\lambda_{r-1} - \tau(\lambda_{r-1})\lambda_r)T_G(\beta_{r-1}) = 0.$$

Since there are fewer than r terms in the relationship, it follows from the minimality of r that all the coefficients must be zero:

$$\tau(\lambda_r)\lambda_i = \tau(\lambda_i)\lambda_r \text{ for } 1 \leqslant i < r;$$

in other words,

$$\tau(\lambda_r^{-1}\lambda_i) = \lambda_r^{-1}\lambda_i \text{ for } 1 \leqslant i < r.$$

Now this holds for each τ in G, and so $k_i = \lambda_r^{-1}\lambda_i \in K$, for $1 \leqslant i < r$. Multiplying (*) by λ_r^{-1}, we obtain

$$k_1\sigma(\beta_1) + \cdots + k_{r-1}\sigma(\beta_{r-1}) + \sigma(\beta_r) = 0$$

for each σ in G. But as G is a subgroup of Aut L, the identity automorphism is in G. Thus

$$k_1\beta_1 + \cdots + k_{r-1}\beta_{r-1} + \beta_r = 0$$

and so B is not linearly independent over K. Thus (i) implies (iii).   □

9.3 The Size of a Galois Group is the Degree of the Extension

When G is finite we can relate the order of G to the degree of $L : \phi(G)$ in a most satisfactory way.

Theorem 9.3 *Suppose that G is a finite subgroup of* Aut L*. Then $|G| = [L : \phi(G)]$, $G = \gamma\phi(G)$ and $L : \phi(G)$ is a Galois extension.*

Proof   Let $K = \phi(G)$. If B is a subset of L which is linearly independent over K then, by Theorem 9.2, $\{T_G(\beta) : \beta \in B\}$ is a subset of L^G which is linearly independent over L. But L^G has dimension $|G|$, and so $|B| \leqslant |G|$. Thus L is finite dimensional over K, and $[L : K] \leqslant |G|$. On the other hand, by Theorem 8.4, $|\gamma\phi(G)| \leqslant [L : K]$. As $G \subseteq \gamma\phi(G)$, it follows that $[L : K] = |G|$ and that $G = \gamma\phi(G)$. Since $[L : K] = |G|$, it follows from Theorem 8.4 that $L : K$ is a Galois extension.   □

What happens if, instead of starting with a group of automorphisms, we start with a finite extension? Here the results are not quite so clear cut. Once again, Theorem 8.4 plays a decisive role.

Theorem 9.4 *Suppose that $L : K$ is finite. If $L : K$ is a Galois extension, then $|\gamma(K)| = [L : K]$, and $K = \phi\gamma(K)$. Otherwise, $|\gamma(K)| < [L : K]$ and K is a proper subfield of $\phi\gamma(K)$.*

Proof The relationship between $|\gamma(K)|$ and $[L : K]$ is given by Theorem 8.4. By Theorem 9.3, $|\gamma(K)| = [L : \phi\gamma(K)]$. Thus, if $L : K$ is normal and separable,

$$[L : K] = [L : \phi\gamma(K)];$$

as $K \subseteq \phi\gamma(K)$, $K = \phi\gamma(K)$. Otherwise

$$[L : K] > [L : \phi\gamma(K)]$$

so that K is a proper subfield of $\phi\gamma(K)$.   □

Exercises

9.1   Suppose that $L : K$ is a Galois extension with Galois group G, and that $\alpha \in L$. Show that $L = K(\alpha)$ if and only if the images of α under G are all distinct.

9.2   Suppose that $L : K$ is an extension. If $\sigma \in \Gamma(L : K)$, $\sigma \in \mathrm{End}_K(L)$, the K-linear space of K-linear mappings of L into itself. Show that $\Gamma(L : K)$ is a linearly independent subset of $\mathrm{End}_K(L)$.

9.3   Suppose that $L : K$ is a Galois extension with Galois group $G = \{\sigma_1, \dots, \sigma_n\}$. Show that $(\beta_1, \dots, \beta_n)$ is a basis for L over K if and only if $\det(\sigma_i(\beta_j)) \neq 0$.

9.4   Suppose that char $K = 0$ and that $L : K$ is a finite extension; let $\beta_1, \dots, \beta_n$ be a basis for L over K. Suppose that H is a subgroup of $\Gamma(L : K)$; let $\gamma_j = \sum_{\sigma \in H} \sigma\beta_j$, for $1 \leqslant j \leqslant n$. Show that $K(\gamma_1, \dots, \gamma_n)$ is the fixed field for H.

9.4 The Galois Group of a Polynomial

The main purpose of the theory of field extensions is to deal with polynomials and their splitting fields.

Suppose that $f \in K[x]$ and that $L : K$ is a splitting field extension for f over K. Then we call $\Gamma(L : K)$ the *Galois group* of f; we denote it by $\Gamma_K(f)$ (or $\Gamma(f)$, when it is clear what K is). By Corollary 1 to Theorem 6.5, $\Gamma_K(f)$ depends on f and K, but not on any particular choice of splitting field.

Let us interpret Theorem 9.4 in this setting.

Theorem 9.5 *Suppose that $f \in K[x]$ and that $L : K$ is a splitting field extension for f. If f is separable then $|\Gamma(f)| = [L : K]$ and $K = \phi(\Gamma(f))$; otherwise $|\Gamma(f)| < [L : K]$ and K is a proper subfield of $\phi(\Gamma(f))$.*

An element σ of $\Gamma(f)$ is an automorphism of L; it is the action of σ on the roots of f that is all important. The next result shows that we lose no information if we concentrate on this action.

Theorem 9.6 *Suppose that $f \in K[x]$ and that $L : K$ is a splitting field extension for f over K. Let R denote the set of roots of f in L. Each σ in $\Gamma(f)$ defines a permutation of R, so that we have a mapping from $\Gamma(f)$ into the group Σ_R of permutations of R. This mapping is a group homomorphism, and is one-to-one.*

Proof If $\sigma \in \Gamma(f)$, then $\sigma(f) = f$, since f has its coefficients in K. Thus, if $\alpha \in R$,

$$f(\sigma(\alpha)) = \sigma(f)(\sigma(\alpha)) = \sigma(f(\alpha)) = \sigma(0) = 0.$$

Thus σ maps R into R. Since σ is one-to-one and R is finite, $\sigma|_R$ is a permutation. By definition,

$$(\sigma_1 \sigma_2)(\alpha) = \sigma_1(\sigma_2(\alpha))$$

so that the mapping: $\sigma \rightarrow \sigma|_R$ is a group homomorphism. Finally, if $\sigma(\alpha) = \tau(\alpha)$ for each α in R, then $\sigma^{-1}\tau$ fixes $K(R) = L$, so that $\sigma = \tau$.

Notice that Theorem 6.6 states that, if f is irreducible, then $\Gamma(f)$ acts *transitively* on the roots of f: if α and β are two roots of f in a splitting field, there exists σ in $\Gamma(f)$ with $\sigma(\alpha) = \beta$.

Conversely, suppose that f is a monic polynomial of degree n in $K[x]$ which has n distinct roots in a splitting field L, and that $\Gamma(f)$ acts transitively on the roots of f Let α be a root of f, and let m be the minimal polynomial of α. Then if β is any root of f there exists σ in $\Gamma(f)$ such that $\sigma(\alpha) = \beta$. Thus

$$m(\beta) = m(\sigma(\alpha)) = \sigma(m)(\sigma(\alpha)) = \sigma(m(\alpha)) = 0,$$

and so m has at least n roots. Since m divides f, $m = f$ and it follows that f is irreducible.   □

Exercises

9.5   Describe the transitive subgroups of Σ_3, Σ_4 and Σ_5.

9.6   Find the Galois group of $x^4 - 2$ over (a) the rational field $\mathbb{Q}$, (b) the field $\mathbb{Z}_3$ and (c) the field $\mathbb{Z}_7$.

9.7   Find the Galois group of $x^4 + 2$ over (a) the rational field $\mathbb{Q}$, (b) the field
$\mathbb{Z}_3$ and (c) the field $\mathbb{Z}_5$.

Let us give an example.

Theorem 9.7 *Suppose that $f \in \mathbb{Q}[x]$ is irreducible and has prime degree p. If
f has exactly $p - 2$ real roots and 2 complex roots in $\mathbb{C}$ then the Galois group
$\Gamma(f)$ of f over $\mathbb{Q}$ is Σ_p.*

Proof For $\Gamma(f)$ is transitive and contains the transposition $z \to \bar{z}$, and so the
result follows from Theorem 1.20.   □

As a concrete example, let us consider

$$f = x^5 - 4x + 2.$$

f is irreducible over $\mathbb{Q}$, by Eisenstein's criterion. The function $t \to f(t)$ on $\mathbb{R}$
is continuous and differentiable, and so, by Rolle's theorem, between any two
real zeros of f there is a zero of f'. But

$$f' = 5x^4 - 4$$

has only two real zeros, so that f has at most three real zeros. As

$$f(-2) = -22, \quad f(0) = 2, \quad f(1) = -1, \quad f(2) = 26$$

f has at least three real roots, by the intermediate value theorem. Thus f has
three real roots and two complex roots; by the theorem, $\Gamma(f) = \Sigma_5$. Notice
how useful elementary analysis can be!

Exercise

9.8   Sketch the graph of the polynomial

$$f_r = (x^2 + 4)x(x^2 - 4)(x^2 - 16)\dots(x^2 - 4r^2).$$

Show that if k is an odd integer then $|f_r(k)| \geqslant 5$. Show that $f_r - 2$
is irreducible, and determine its Galois group over $\mathbb{Q}$ when $2r + 3$ is a
prime.

9.5 The Fundamental Theorem of Galois Theory

The fundamental theorem of Galois theory describes in some detail the polarity
that was introduced at the beginning of the chapter.

Theorem 9.8 *Suppose that $L : K$ is finite. Let $G = \Gamma(L : K)$, and let $K_0 =
\phi(G)$. If $L : M : K_0$, let $\gamma(M) = \Gamma(L : M)$.*

(i) *The map ϕ is a one-to-one map from the set of subgroups of G onto the set of fields M intermediate between L and K_0. γ is the inverse map.*

(ii) *A subgroup H of G is normal if and only if $\phi(H) : K_0$ is a normal extension.*

(iii) *Suppose that $H \lhd G$. If $\sigma \in G$, $\sigma|_{\phi(H)} \in \Gamma(\phi(H) : K_0)$. The map $\sigma \to \sigma|_{\phi(H)}$ is a homomorphism of G onto $\Gamma(\phi(H) : K_0))$, with kernel H. Thus*

$$\Gamma(\phi(H) : K_0) \cong G/H.$$

Proof   (i) If H is a subgroup of G, H is finite, and so $\gamma\phi(H) = H$ (Theorem 9.3). Thus ϕ is one-to-one. $L : K_0$ is a Galois extension of K_0 (Theorem 9.3); thus if $L : M : K_0$, $L : M$ is normal (Corollary 1 of Theorem 7.1) and separable (Theorem 8.1). By Theorem 9.4, $\phi\gamma(M) = M$. Thus ϕ is onto, and γ is the inverse mapping.

  (ii) Suppose that $L : M : K_0$, and that $\sigma \in G$. Then $L : \sigma(M) : K_0$ (since $\sigma(L) = L$, $\sigma(K_0) = K_0$). We shall show that $\gamma(\sigma(M)) = \sigma(\gamma(M))\sigma^{-1}$. For $\tau \in \gamma(\sigma(M))$ if and only if $\tau\sigma(m) = \sigma(m)$ for each m in M, and this happens if and only if $\sigma^{-1}\tau\sigma(m) = m$ for each m in M. Thus $\tau \in \gamma(\sigma(M))$ if and only if $\sigma^{-1}\tau\sigma \in \gamma(M)$: that is, if and only if $\tau \in \sigma(\gamma(M))\sigma^{-1}$.

  Suppose that $H \lhd G$. Then, if $\sigma \in G$, using (i)

$$H = \sigma H \sigma^{-1} = \sigma(\gamma\phi(H))\sigma^{-1} = \gamma(\sigma(\phi(H))).$$

Applying ϕ, $\phi(H) = \phi\gamma(\sigma(\phi(H))) = \sigma(\phi(H))$ for each σ in G. By Theorem 7.2, $\phi(H) : K_0$ is normal.

  Conversely if $\phi(H) : K_0$ is normal, then

$$\phi(H) = \sigma\phi(H) \text{ for each } \sigma \text{ in } G,$$

by Theorem 7.2. Thus

$$H = \gamma\phi(H) = \gamma(\sigma(\phi(H))) = \sigma(\gamma\phi(H))\sigma^{-1} = \sigma H \sigma^{-1}$$

for each σ in G, and so $H \lhd G$.

  (iii) Suppose now that $H \lhd G$, so that $\phi(H) : K_0$ is normal. If $\sigma \in G$, $\sigma(\phi(H)) = \phi(H)$, by Theorem 7.2. Thus $\sigma|_{\phi(H)}$ is an automorphism of $\phi(H)$ fixing K_0: that is, an element of $\Gamma(\phi(H) : K_0)$. Since the group multiplication is the composition of mappings, the mapping $\sigma \to \sigma|_{\phi(H)}$ is a homomorphism of G into $\Gamma(\phi(H) : K_0)$. σ is the kernel of this homomorphism if and only if $\sigma|_{\phi(H)}$ is the identity: that is, if and only if σ fixes $\phi(H)$. Thus the kernel is $\gamma\phi(H) = H$. Finally, if $\rho \in \Gamma(\phi(H) : K_0)$, there exists a monomorphism $\sigma : L \to L$ which extends ρ (Theorem 8.3). As $[L : \phi(H)] = [\sigma(L) : \sigma\phi(H)] =$

$[\sigma(L) : \phi(H)]$, $\sigma(L) = L$ and σ is an automorphism. As ρ fixes K_0, σ is in G; as $\sigma|_{\phi(H)} = \rho$, the homomorphism maps G onto $\Gamma(\phi(H) : K_0)$. $\qquad\square$

This completes the proof. A few remarks are in order. First, the proof is, to a large extent, a case of putting together results which have been established earlier. It is worth pausing, and tracing these results back to their sources: this frequently turns out to be Theorem 6.4. Second, the theorem relates subgroups of G to intermediate fields: order is reversed and, the smaller the subgroup, the larger is the intermediate field. *All* the subgroups of G relate to *all* the intermediate fields M of $L : K_0$. Remember that this occurs in terms of $\Gamma(L : M)$, and *not* $\Gamma(M : K_0)$. *Normal* subgroups relate to intermediate fields M for which $M : K_0$ is *normal*; this justifies the terminology. If $M : K_0$ is normal, we can calculate $\Gamma(M : K_0)$ in terms of $\Gamma(L : K)$, but as a *quotient*, not as a *sub*group. Finally, we do not need normality or separability. But if $L : K$ is a Galois extension, then $K = K_0$, and the result is correspondingly neater.

Exercises

9.9   Given a finite group G show that there exists a Galois extension $L : K$ such that $\Gamma(L : K) \cong G$.

9.10 Suppose that K_1 and K_2 are subfields of a field L such that $L : K_1$ and $L : K_2$ are both Galois extensions, with Galois groups G_1 and G_2 respectively. Show that $L : K_1 \cap K_2$ is a Galois extension if and only if G, the group generated by G_1 and G_2, is finite, and that if this is so then $G = \Gamma(L : K_1 \cap K_2)$.

9.11 Suppose that $L : K$ is an extension with $[L : K] = 2$, that every element of L has a square root in L, that every polynomial of odd degree in $K[x]$ has a root in K and that char $K \neq 2$. Let f be an irreducible polynomial in $K[x]$, let $M : L$ be a splitting field extension for f over L, let $G = \Gamma(M : K)$ and let $H = \Gamma(M : L)$.

  (i)   By considering the fixed field of a Sylow 2-subgroup of G, show that $|G| = 2^n$.

  (ii) By considering a subgroup of index 2 in H, show that if $n > 1$ then there is an irreducible quadratic in $L[x]$.

  (iii) Show that L is algebraically closed.

  (iv) Show that the complex numbers are algebraically closed.

9.12 By considering the splitting field of all polynomials of odd degree over $\mathbb{Z}_2$, show that the condition char $K \neq 2$ cannot be dropped from Exercise 9.11.

9.13 Suppose that $L : K$ is a Galois extension with Galois group $G = (g_1, \ldots, g_n)$. Show that $(\alpha_1, \ldots, \alpha_n)$ is a basis for $L : k$ if and only if $det(g_i(\alpha_j)) \neq 0$.

9.14 Suppose that $L : K$ is a finite Galois extension, that M and M' are intermediate fields, and that $\phi : M \to M'$ is a monomorphism which fixes the points of K. Show that there is σ in the Galois group of $L : K$ whose restriction to M is ϕ.

10

The Discriminant

10.1 The Discriminant

Suppose that K is a field whose characteristic is not 2, that f is a normal separable monic polynomial of degree n in $K[x]$ and that $L : K$ is a splitting field extension for f. Then the Galois group $\Gamma[L : K]$ acts on the set $\{\alpha_1, \ldots, \alpha_n\}$ of roots of f in L, and can therefore be identified with a subgroup of Σ_n. When is $\Gamma[L : K]$ contained in A_n? If it is not, can we extend K to $K(\delta)$ in such a way that $\Gamma[L : K(\delta)] = \Gamma[L : K] \cap A_n$? These questions can be answered by introducing the *discriminant* of f.

Let $\delta = \prod_{i<j}(\alpha_i - \alpha_j)$. Note that there is ambiguity in this definition, since it depends upon the order in which the roots are labelled. There is however no ambiguity in the discriminant $\Delta = \delta^2 = \prod_{i<j}(\alpha_i - \alpha_j)^2$. We have defined Δ as an element of L; in fact, it belongs to K.

Theorem 10.1 *Suppose that f is a normal separable polynomial in $K(x)$, that $L : K$ is a splitting field extension and that Δ is the discriminant of f.*

 (i) *The discriminant Δ of f is an element of K.*
 (ii) *If $\Delta = 0$, then f has a repeated root.*
 (iii) *If $\Delta \neq 0$, then Δ is a square in K if and only if $\Gamma_K(f) \subseteq A_n$.*
 (iv) *If Δ is not a square in K, then it has a square root δ in L, $[K(\delta) : K] = 2$ and $\Gamma_{K(\delta)}(f) = \Gamma_K(f) \cap A_n$.*

Proof   (i) If $\sigma \in \Gamma_K(f)$, then $\sigma(\Delta) = \prod_{i<j}(\alpha_{\sigma(i)} - \alpha_{\sigma(j)})^2 = \Delta$, so that Δ is in the fixed field of $\Gamma_K(f)$, which is K.

 (ii) This follows from the definition of Δ.

 (iii) If Δ is a square in K, then $\delta \in K$, so that if $\sigma \in \Gamma_K(f)$, then $\sigma(\delta) = \prod_{i<j}(\alpha_{\sigma(i)} - \alpha_{\sigma(j)}) = \delta$ and $\sigma \in A_n$. Conversely, if $\Gamma_K(f) \subseteq A_n$ then $\sigma(\delta) = \delta$ for each σ in $\Gamma_K(f)$, so that $\delta \in K$.

(iv) If Δ is not a square in K, then $K(\delta) : K$ is a splitting field extension for $x^2 - \Delta$ and $[K(\delta) : K] = 2$. Now $\Gamma_K(f) \cap A_n$ has index 2 in $\Gamma_K(f)$ and fixes $K(\delta)$, and so it follows from the fundamental theorem of Galois theory that $K(\delta)$ is the fixed field of $\Gamma_K(f) \cap A_n$ and that $\Gamma_K(f) \cap A_n = \Gamma_{K(\delta)}(f)$. $\square$

But how do we calculate the discriminant, which has been defined in terms of the roots, which we do not know. The quantity δ is given by the Vandermonde determinant

$$\delta = \begin{vmatrix} 1 & 1 & \cdots & 1 \\ \alpha_1 & \alpha_2 & \cdots & \alpha_n \\ \vdots & \vdots & \ddots & \vdots \\ \alpha_1^{n-1} & \alpha_2^{n-1} & \cdots & \alpha_n^{n-1} \end{vmatrix}$$

If we multiply the matrix by its transpose and calculate the determinant, we find that

$$\Delta = \begin{vmatrix} n & \lambda_1 & \cdots & \lambda_{n-1} \\ \lambda_1 & \lambda_2 & \cdots & \lambda_n \\ \vdots & \vdots & \ddots & \vdots \\ \lambda_{n-1} & \lambda_n & \cdots & \lambda_{2n-1} \end{vmatrix}$$

where $\lambda_j = \alpha_1^j + \cdots + \alpha_n^j$. Now we appeal to the Girard–Newton formula.

Theorem 10.2 *Suppose that $\{x_1, \ldots, x_n\}$ are elements of a field K, that $\sigma_k = \sum_{|A|=k} (\prod_{x_i \in A} x_i)$ and that $\lambda_i = \sum_{j=1}^n x_j^i$, where $k \geq 0$. (Thus $\lambda_0 = n$; we put $\sigma_0 = 1$.) Then $\sum_{l=0}^k (-1)^l \sigma_l \lambda_{k-l} = 0$ for $1 \leq k \leq n$.*

Proof Let $S = \{x_1, \ldots, x_n\}$ and let $\mathcal{B} = P(S) \times S$. If $B = (A, x_j) \in \mathcal{B}$ let $d(B) = n - |A|$ and let $v(B) = (-1)^{|A|} (\prod_{x_i \in A} x_i) . x_j^{d(A)}$. Then $\sum_{l=0}^k (-1)^l \sigma_l \lambda_{k-l} = \sum_\mathcal{B} v(B)$. If $(A, j) \in \mathcal{B}$ let $T(A, x_j) = (A \backslash \{x_j\}, x_j)$ if $x_j \in A$, and $T(A, x_j) = (A \cup \{x_j\}, x_j)$ if $x_j j \notin A$. Then T is a bijection of $\mathcal{B}$ to itself. Since $v(T(B)) = -v(B)$ for each $B \in \mathcal{B}$, $\sum_\mathcal{B} v(B) = 0$. $\square$

Corollary 10.3 *If $\alpha_1, \ldots, \alpha_n$ are roots of the polynomial equation $x_n + a_{n-1}x^{n-1} + \cdots + a_0 = 0$ then $\sum_{l=0}^k a_l \lambda_l = 0$ for $1 \leq k \leq n$.*

Proof For $a_l - (-1)^l \sigma n - l$. $\square$

Thus Δ can be computed: tedious with pen and paper, but quite reasonable on a computer.

Here are some examples.

If $f = x^2 + bx + c$ then $\Delta = b^2 - 4c$.

If $f = x^3 + px + q$ then $\Delta = -4p^3 - 27q^2$.
If $f(x) = x^4 + px^2 + qx + r$ then

$$\Delta = 16p^4 r - 4p^3 q^2 - 128 p^2 r^2 + 144 pq^2 r - 27q^4 + 256 r^3.$$

Exercises

10.1 Suppose that f is a polynomial in $K[x]$, with roots $\alpha_1, \ldots, \alpha_n$ in some splitting field extension. Show that

$$\Delta = \eta_n \prod_{j=1}^{n} Df(\alpha_j),$$

where $\eta_n = 1$ if $n \pmod 4 = 0$ or 1 and $\eta_n = -1$ otherwise.

10.2 Suppose that

$$f = a_0 + a_1 x + \cdots + a_n x^n$$

is a polynomial of degree n in $K[x]$ and that $\alpha_1, \ldots, \alpha_n$ are roots of f in a splitting field L.

(i) Show that, in $L[x]$, $f = (x - \alpha_i)g_i$, where

$$g_i = a_1 + a_2(x + \alpha_i) + \cdots + a_n(x^{n-1} + \alpha_i x^{n-2} + \cdots + \alpha_i^{n-1}).$$

(ii) Show that $Df = \sum_{i=1}^{n} g_i$.

(iii) Let $\lambda_j = \sum_{i=1}^{n} \alpha_i^j$, for $j = 1, 2, \ldots$ Establish *Newton's identities*:

$$a_{n-1} + a_n \lambda_1 = 0,$$
$$2a_{n-2} + a_{n-1}\lambda_1 + a_n \lambda_2 = 0,$$
$$\vdots$$
$$na_0 + a_1\lambda_1 + \cdots + a_{n-1}\lambda_{n-1} + a_n\lambda_n = 0,$$

and

$$a_0\lambda_k + a_1\lambda_{k+1} + \cdots + a_{n-1}\lambda_{k+n-1} + a_n\lambda_{k+n} = 0$$

for $k = 1, 2, 3, \ldots$

10.3 Suppose that $f = x^n + px + q$. Show that

$$\lambda_1 = \lambda_2 = \cdots = \lambda_{n-2} = 0,$$
$$\lambda_{n-1} = -(n-1)p,$$
$$\lambda_n = -nq,$$
$$\lambda_{n+1} = \cdots = \lambda_{2n-3} = 0$$

and

$$\lambda_{2n-2} = (n-1)p^2.$$

Show that the discriminant Δ of f is

$$\Delta = \eta_{n+1} n^n q^{n-1} - \eta_n (n-1)^{n-1} p^n$$

where $\eta_n = 1$ if $n \pmod 4 = 0$ or 1 and $\eta_n = -1$ otherwise.

10.4 Suppose that char $K = 2$ and that $f \in K[x]$ is separable. Show that the discriminant of f always has a square root in K.

10.5 Suppose that $f \in \mathbb{Q}[x]$ is irreducible, and that its discriminant δ does not have a square root in $\mathbb{Q}$. If δ is a square root of $\wedge$ in some splitting field, show that f is irreducible in $\mathbb{Q}(\delta)$.

11

Cyclotomic Polynomials and Cyclic Extensions

We have seen that in order to deal with cubic polynomials it is helpful to have cube roots of unity at our disposal. In this chapter and the next we shall consider splitting fields and Galois groups of polynomials of the form $x^m - 1$ and $x^m - \theta$ over a field K.

Technical problems can arise if char $K \neq 0$. Suppose that char $K = p > 0$ and that $m = p^r q$, where p does not divide q. Then in $K[x]$,

$$x^m - 1 = (x^q - 1)^{p^r};$$

thus a splitting field extension for $x^q - 1$ is a splitting field extension for $x^m - 1$: we need only consider the polynomial $x^q - 1$. For this reason, *in this chapter we shall suppose that* char K *does not divide* m. In this case, $D(x^m - 1) = mx^{m-1} \neq 0$, and so $x^m - 1$ has m distinct roots in a splitting field.

11.1 Cyclotomic Polynomials

Suppose that $L : K$ is a splitting field extension for $x^m - 1$ over K. As $x^m - 1$ has m distinct roots, $L : K$ is a Galois extension. The set R of roots in L clearly forms a group under multiplication, and so, by Exercise 10.1, R is a cyclic group of order m. An element ε of R is called a *primitive mth root of unity* if ε generates R. Thus an element ε of L is a primitive mth root of unity if and only if $\varepsilon^m = 1$ and $\varepsilon^j \neq 1$ for $1 \leqslant j < m$. For example, in $\mathbb{C}$, i and $-i$ are the primitive fourth roots of unity: -1 is the only primitive second root of unity and 1 is the only first root of unity. Notice that if ε is a primitive mth root of unity then $L = K(\varepsilon)$.

We now define the *mth cyclotomic polynomial* Φ_m to be

$$\Phi_m = \prod_\varepsilon (x - \varepsilon)$$

where the product is taken over all *primitive* mth roots of unity. An element α in L is a root of $x^m - 1$ if and only if it is a primitive dth root of unity for some d which divides m: thus

$$x^m - 1 = \prod_{d|m} \Phi_d.$$

For example, in $\mathbb{C}[x]$

$$\Phi_1 = x - 1, \quad \Phi_3 = (x - \omega)(x - \omega^2) = x^2 + x + 1$$
$$\Phi_2 = x + 1, \quad \Phi_4 = (x - i)(x + i) = x^2 + 1$$

and

$$x^4 - 1 = (x - 1)(x + 1)(x^2 + 1) = \Phi_1 \Phi_2 \Phi_4.$$

We have defined Φ_m as an element of $L[x]$. In fact, as the examples suggest, we can say much more.

Theorem 11.1 $\Phi_m \in K_0[x]$, where K_0 is the prime subfield of K. If $K_0 = \mathbb{Q}$ then $\Phi_m \in \mathbb{Z}[x]$.

Proof Since $x^m - 1 = \prod_{d|m} \Phi_d$, the theorem follows from an inductive application of the following elementary lemma. □

Lemma 11.2 (i) *If $L: K$ is an extension, if $q \in L[x]$ and if there exist non-zero f and g in $K[x]$ such that $f = qg$, then $q \in K[x]$.*

(ii) *Suppose that K is the field of fractions of an integral domain R, that $q \in K[x]$ and that there exist monic f and g in $R[x]$ such that $f = qg$. Then $q \in R[x]$.*

Proof This is just a matter of long division.

(i) Let

$$q = a_0 + a_1 x + \cdots + a_m x^m,$$
$$g = b_0 + b_1 x + \cdots + b_n x^n,$$
$$f = c_0 + c_1 x + \cdots + c_{m+n} x^{m+n}$$

where $a_m \neq 0$, $b_n \neq 0$, $c_{m+n} \neq 0$. As $a_m b_n = c_{m+n}$, $a_m \in K$. Suppose that we have shown that $a_i \in K$ for $i > j$. Then as

$$a_j b_n + a_{j+1} b_{n-1} + \cdots + a_m b_{n+j-m} = c_{n+j}$$

(where we set $b_k = 0$ if $k < 0$), $a_j \in K$.

(ii) In this case $b_n = 1$, and the same induction goes through. □

Exercises

11.1 Show that the degree of Φ_m is $m \prod_{p|m} ((p-1)/p)$, where the product is taken over all primes p which divide m.

11.2 Show that if n is odd and $n > 1$ then $\Phi_{2n}(x) = \Phi_n(-x)$.

11.3 Show that if p is a prime then
$$\Phi_{p^m}(x) = 1 + x^{p^{n-1}} + x^{2p^{n-1}} + \cdots + x^{(p-1)p^{n-1}}.$$

11.4 If p is a prime number, show that $\Phi_{np^2}(x) = \Phi_{np)}(x^p)$.

11.5 Suppose that p, q are different prime numbers. Show that the coefficients of Φ_{pq} are alternately 1 and -1.

11.6 Calculate Φ_n for $n = 18, 24$ and 30.

11.2 Irreducibility

By Theorem 11.1, we can consider the cyclotomic polynomials Φ_m as polynomials in $K_0[x]$, where K_0 is the prime subfield of K. In the case where char $K_0 \neq 0$, the irreducibility of Φ_m over K_0 depends upon m and char K_0: for example, $\Phi_3 = x^2 + x + \bar{1}$ is irreducible over $\mathbb{Z}_5$, while over $\mathbb{Z}_7$
$$x^2 + x + \bar{1} = (x - \bar{2})(x - \bar{4}).$$

In the important case where char $K_0 = 0$, the result is simple to state, but remarkably difficult to prove:

Theorem 11.3 *For each m, Φ_m is irreducible over $\mathbb{Q}$.*

Proof Suppose that Φ_m is not irreducible. By Gauss' lemma we can write $\Phi_m = fg$, where f and g are in $\mathbb{Z}[x]$ and f is an irreducible monic polynomial with $1 \leqslant$ degree $f <$ degree Φ_m.

Let $L : \mathbb{Q}$ be a splitting field extension for Φ_m over $\mathbb{Q}$. We shall first show that, if ε is a root of f in L and p is a prime which does not divide m, then ε^p is a root of f.

Suppose not. Then, as ε^p is a primitive mth root of unity, $g(\varepsilon^p) = 0$. We define k in $\mathbb{Z}[x]$ by setting $k(x) = g(x^p)$. Then $k(\varepsilon) = g(\varepsilon^p) = 0$. Since f is the minimal polynomial for ε over $\mathbb{Q}$, $f|k$ in $\mathbb{Q}[x]$, and, by Lemma 11.2, we can write $k = fh$, with h in $\mathbb{Z}[x]$.

We now consider the quotient map: $n \to \bar{n}$ from $\mathbb{Z}$ onto $\mathbb{Z}_p$, and the induced map: $j \to \bar{j}$ of $\mathbb{Z}[x]$ onto $\mathbb{Z}_p[x]$. Under this map, $\overline{fh} = \bar{k}$. But
$$\overline{k(x)} = \overline{g(x^p)} = (\overline{g(x)})^p$$

and so $\overline{f}\,\overline{h} = (\overline{g})^p$. Let $\overline{q}$ be any irreducible factor of $\overline{f}$ in $\mathbb{Z}_p[x]$. Then $\overline{q}\,|\,(\overline{g})^p$, and $\overline{q}\,|\,\overline{g}$. This means that $\overline{q}^2\,|\,\overline{f}\,\overline{g}$, so that $\overline{\Phi}_m = \overline{f}\,\overline{g}$ has a repeated root in a splitting field extension over $\mathbb{Z}_p$. But we have seen that this is not so, since p dooe not divide m.

Now let η be a root of f, and let θ be a root of g. θ and η are both primitive mth roots of unity, and so there exists r such that $\theta = \eta^r$, where r and m are relatively prime. We can write $r = p_1 \ldots p_k$ as a product of primes, where no p_i divides m. Repeated application of the result that we have proved shows that θ is a root of f. This means that Φ_m has a repeated root in L, and we know that this is not so. □

Exercise

11.7 Suppose that ε is a primitive mth root of unity over $\mathbb{Q}$, where $m > 2$. Let $\eta = \varepsilon + \varepsilon^{-1}$. Show that $[\mathbb{Q}(\varepsilon) : \mathbb{Q}(\eta)] = 2$, find the minimal polynomial for ε over $\mathbb{Q}(\eta)$ and identify the Galois group $\Gamma[\mathbb{Q}(\varepsilon) : \mathbb{Q}(\eta)]$.

11.3 The Galois Group of a Cyclotomic Polynomial

Suppose that $L : K$ is a splitting field extension for the cyclotomic polynomial Φ_m over K. If ε is a primitive mth root of unity then, as we have seen, $L = K(\varepsilon)$.

We can write the primitive mth roots of unity as

$$\varepsilon^{n_1}, \varepsilon^{n_2}, \ldots, \varepsilon^{n_k},$$

where $1 = n_1, n_2, \ldots, n_k$ are those integers less than m which are relatively prime to m, and $k = $ degree Φ_m. Now, if n_i and m are relatively prime, $\mathbb{Z} = (n_i, m)$ and so there exist integers a and b such that $an_i + bm = 1$. Thus in the quotient ring $\mathbb{Z}_m$, $\overline{a}\,\overline{n}_i = \overline{1}$, and n_i is a unit. Conversely if $\overline{n}$ is a unit in $\mathbb{Z}_m$ then n and m are relatively prime. Thus $\{\overline{n}_1, \ldots, \overline{n}_k\}$ is the *multiplicative* group U_m of units in the ring $\mathbb{Z}_m$.

Now suppose that σ is in the Galois group $\Gamma_K(\Phi_m)$. As $L = K(\varepsilon)$, σ is determined by its action on ε. As $\sigma(\varepsilon)$ is also a primitive mth root of unity, $\sigma(\varepsilon) = \varepsilon^{n_{j(\sigma)}}$ for some $1 \leqslant j(\sigma) \leqslant k$. If τ is another element of $\Gamma_K(\Phi_m)$,

$$\tau\sigma(\varepsilon) = \tau(\varepsilon^{n_{j(\sigma)}}) = (\tau(\varepsilon))^{n_{j(\sigma)}} = \varepsilon^{n_{j(\tau)}n_{j(\sigma)}} = \sigma\tau(\varepsilon).$$

Thus $\Gamma_K(\Phi_m)$ is abelian. Also

$$\overline{n}_{j(\tau\sigma)} = \overline{n}_{j(\tau)}\overline{n}_{j(\sigma)}$$

and so the mapping $\sigma \to \bar{n}_{j(\sigma)}$ is a homomorphism of $\Gamma_K(\Phi_m)$ into U_m. This is injective, since $\sigma(\varepsilon) = \varepsilon$ if and only if σ is the identity in $\Gamma_K(\Phi_m)$. Further, $|\Gamma_K(\Phi_m)| = k$ if and only if there are k images $\varepsilon^{n_{j(\sigma)}}$; thus the homomorphism is onto if and only if $\Gamma_K(\Phi_m)$ acts transitively on the roots of Φ_m, and this happens if and only if Φ_m is irreducible over K. Summing up:

Theorem 11.4 *If Φ_m is the mth cyclotomic polynomial over K, $\Gamma_K(\Phi_m)$ is an abelian group which is isomorphic to a subgroup of U_m, the multiplicative group of the ring $\mathbb{Z}_m$. Φ_m is irreducible over K if and only if $\Gamma_K(\Phi_m)$ is isomorphic to U_m.*

As an example,

$$U_{12} = \{\bar{1}, \bar{5}, \bar{7}, \overline{11}\}$$

and $\bar{1}^2 = \bar{5}^2 = \overline{11}^2 = \bar{1}$, so that $U_{12} \cong \mathbb{Z}_2 \times \mathbb{Z}_2$.

If p is a prime, U_p is cyclic, by Exercise 10.1. We therefore have the following corollary:

Corollary 1 *If p is a prime then either Φ_p splits over K or $\Gamma_K(\Phi_p)$ is cyclic.*

Corollary 2 *For each m, $\Gamma_{\mathbb{Q}}(\Phi_m)$ is isomorphic to U_m.*

Exercises

11.8   Find the Galois groups of $x^4 + 1$ and $x^5 + 1$ over $\mathbb{Q}$.

11.9   Suppose that p is a prime which does not divide m, and let ε be a primitive mth root of unity over $\mathbb{Z}_p$. Show that $[\mathbb{Z}_p(\varepsilon) : \mathbb{Z}_p] = k$, where k is the order of $\bar{p}$ in the multiplicative group U_m of units in $\mathbb{Z}_m$. Show that Φ_m is irreducible over $\mathbb{Z}_p$ if and only if U_m is a cyclic group generated by $\bar{p}$. When is Φ_4 irreducible over $\mathbb{Z}_p$? When is Φ_8 irreducible over $\mathbb{Z}_p$?

11.10 Suppose that $m = q^t$, where q is an odd prime.

(i)   Show that $|U_m| = (q-1)\phi^{t-1}$.

(ii) Use the fact that U_q is cyclic of order $q - 1$ to show that there is an element of order $q - 1$ in U_m.

(iii) Show that if q does not divide a then

$$(1 + aq^u)^q = 1 + bq^{u+1},$$

where q does not divide b.

(iv) Show that $\overline{1+q}$ has order q^{t-1} in U_m.

(v)   Combine (ii) and (iv) to show that U_m is cyclic.

11.11 Suppose that $m = m_1 \ldots m_r$, where $m_1, \ldots, m_r$ are distinct prime powers. Show that

$$U_m \cong U_{m_1} \times \cdots \times U_{m_r}.$$

11.12 Show that U_m is cyclic if and only if $m = q^t$ or $2q^t$ (where q is an odd prime) or 4.

11.13 Is Φ_{18} irreducible over (a) $\mathbb{Z}_{23}$, (b) $\mathbb{Z}_{43}$, (c) $\mathbb{Z}_{73}$?

If we are going to study extensions by radicals, it is clearly useful to be able to answer the question: when is a Galois extension $L : K$ a splitting field extension for a polynomial of the form $x^n - \theta$?

11.4 A Necessary Condition

We begin by considering a polynomial of the form $x^n - \theta$ in $K[x]$. In order to avoid problems of separability, let us suppose that char K does not divide n. Let $L : K$ be a splitting field extension for $f = x^n - \theta$ over K. Then by Theorem 8.5, f has n distinct roots, $\alpha_1, \ldots, \alpha_n$ say, in L. Since $(\alpha_i \alpha_j^{-1})^n = \theta\theta^{-1} = 1$, the elements $\alpha_1\alpha_1^{-1}, \alpha_2\alpha_1^{-1}, \ldots, \alpha_n\alpha_1^{-1}$ are n distinct roots of unity in L, so that $x^n - 1$ splits over L. Let ω be a primitive nth root of unity. Then

$$x^n - \theta = (x - \alpha_1)(x - \omega\alpha_1) \ldots (x - \omega^{n-1}\alpha_1).$$

This suggests that we should consider the intermediate field $K(\omega)$, which contains all the nth roots of unity.

A field extension $L : K$ is a *cyclic extension* if it is a Galois extension whose Galois group is cyclic.

Theorem 11.5 *Suppose that $x^n - \theta \in K[x]$, and that char K does not divide n. Let $L : K$ be a splitting field extension for $x^n - \theta$ over K. Then L contains a primitive nth root of unity, ω say. The extension $L : K(\omega)$ is a cyclic extension, and its order divides n. $x^n - \theta$ is irreducible over $K(\omega)$ if and only if $[L : K(\omega)] = n$.*

Proof We have seen that L contains a primitive nth root of unity and that $x^n - \theta$ splits over L as

$$(x - \beta)(x - \omega\beta) \ldots (x - \omega^{n-1}\beta).$$

Thus $L = K(\omega, \beta)$ and if $\sigma \in \Gamma(L : K(\omega))$, σ is determined by its action on β. Now if σ is in $\Gamma(L : K(\omega))$

$$\sigma(\beta) = \omega^{j(\sigma)}\beta$$

for some $0 \leqslant j(\sigma) < n$. Since $\omega \in K(\omega)$, if σ and τ are in $\Gamma(L : K(\omega))$,

$$\tau\sigma(\beta) = \tau(\omega^{j(\sigma)}\beta) = \omega^{j(\sigma)}\tau(\beta) = \omega^{j(\sigma)}\omega^{j(\tau)}\beta$$

and so the map $\sigma \to \overline{j(\sigma)}$ is a homomorphism of $\Gamma(L : K(\omega))$ into the *additive* group $(\mathbb{Z}_n, +)$. As $\overline{j(\sigma)} = \overline{0}$ if and only if $\sigma(\beta) = \beta$, and this happens if and only if σ is the identity, the homomorphism is one-to-one. Thus $\Gamma(L : K(\omega))$ is isomorphic to a subgroup of the cyclic group $(\mathbb{Z}_n, +)$: it is therefore cyclic, and its order divides n.

If $x^n - \theta$ is irreducible over $K(\omega)$, $|\Gamma(L : K(\omega))| \geqslant n$, so that $[L : K(\omega)] = |\Gamma(L : K(\omega))| = n$. If $x^n - \theta$ is not irreducible over $K(\omega)$, let g be an irreducible monic factor in $K(\omega)$, and let γ be a root of g in L. Then

$$x^n - \theta = (x - \gamma)(x - \omega\gamma) \dots (x - \omega^{n-1}\gamma)$$

so that $x^n - \theta$ splits over $K(\omega, \gamma)$, and $L = K(\omega, \gamma)$. Thus

$$[L : K(\omega)] = [K(\omega, \gamma) : K(\omega)] = \text{degree } g < n. \qquad \square$$

Exercises

11.14 Show that $x^6 + 3$ is irreducible over $\mathbb{Q}$, but is not irreducible over $\mathbb{Q}(\omega)$, where ω is a primitive sixth root of unity.

11.15 Show that the Galois group of $x^{15} - 2$ over $\mathbb{Q}$ can be generated by elements ρ, σ and τ satisfying

$$\rho^{15} = \sigma^4 = \tau^2 = 1,$$
$$\sigma^{-1}\rho\sigma = \rho^7,$$
$$\tau^{-1}\rho\tau = \rho^{14},$$
$$\tau^{-1}\sigma\tau = \sigma.$$

11.16 Let $L : \mathbb{Q}$ be a splitting field extension for $x^4 - 5$ over $\mathbb{Q}$. What is its Galois group? List the fields intermediate between L and $\mathbb{Q}$, and determine which of them are normal over $\mathbb{Q}$.

11.5 Abel's Theorem

In the case where n is a prime, we can say more about the irreducibility of $x^n - \theta$.

Theorem 11.6 (Abel's theorem) *Suppose that q is a prime, that $x^q - \theta \in K[x]$ and that* char $K \neq q$. *Then either $x^q - \theta$ is irreducible over K or $x^q - \theta$ has* *a root in K. In the latter case $x^q - \theta$ splits over K if and only if K contains a primitive qth root of unity.*

Proof Suppose that $x^q - 0$ is not irreducible over K. Let $L : K$ be a splitting field extension for $x^q - \theta$, let g be an irreducible monic divisor of $x^q - \theta$ in $K[x]$ and let γ be a root of g in L. Then, in L,

$$g = (x - \gamma)(x - \omega^{n_2}\gamma)\ldots(x - \omega^{n_d}\gamma)$$

where ω is a primitive qth root of unity in L, $1 \leqslant n_2 < n_3 < \cdots < n_d < q$ and $d = $ degree g. Thus if

$$g = x^d - g_{d-1}x^{d-1} + \cdots + (-1)^d g_0,$$

$g_0 = \omega^k \gamma^d$ for some k. Raising this to the qth power, we see that $g_0^q = \gamma^{dq} = \theta^d$. Now d and q are coprime, and so there exist integers a and b such that

$$ad + bq = 1.$$

Thus

$$\theta = \theta^{ad}\theta^{bq} = (g_0^a\theta^b)^q$$

and so $x^q - \theta$ has a root $g_0^a\theta^b$ in K.

If $x^q - \theta$ is not irreducible over K, $[L : K(\omega)]$ divides q and is less than q, by Theorem 11.5, and so $L = K(\omega)$. Thus $K(\omega) : K$ is a splitting field extension for $x^q - \theta$: the last statement of the theorem follows immediately from this. $\square$

Exercises

11.17 Suppose that q is a prime, that char $K \neq q$ and that $x^q - \theta$ is irreducible in $K[x]$. Let ω be a primitive qth root of unity, and let $[K(\omega) : K] = j$. Show that the Galois group of $x^q - \theta$ can be generated by elements σ and τ satisfying

$$\sigma^q = \tau^j = 1, \quad \sigma^k\tau = \tau\sigma,$$

where $\bar{k}$ is a generator of the multiplicative group $\mathbb{Z}_q^*$.

11.18 Suppose that q is a prime, that char $K = q$ and that $\theta \in K$. Describe the splitting field for $x^q - \theta$ over K.

11.6 Norms and Traces

Suppose that $L : K$ is a Galois extension of degree n, with Galois group G. If $\alpha \in L$, let $N_{L:K}(\alpha) = \prod_{\sigma \in G} \sigma(\alpha)$ and $tr_{L:K}(\alpha) = \sum_{\sigma \in G} \sigma(\alpha)$. $N_{L:K}$ is the *norm* of α and $tr_{L:K}(\alpha)$ is the *trace* of α. Here are the elementary properties of the norm and the trace.

Proposition 11.7 *Suppose that $L : K$ is a Galois extension of degree n, with Galois group G and that $\alpha, \beta \in L$.*

(i) $N_{L:K}(\alpha) \in K$ and $tr_{L:K}(\alpha) \in K$.

(ii) *If $\alpha \in K$ then $N_{L:K}(\alpha) = \alpha^n$ and $tr_{L:K}(\alpha) = n\alpha$.*

(iii) $N_{L:K}(\alpha\beta) = N_{L:K}(\alpha)N_{L:K}(\beta)$ and $tr_{L:K}(\alpha+\beta) = tr_{L:K}(\alpha)+tr_{L:K}(\beta)$.

Proof   Since $N_{L:K}(\alpha) = N_{L:K}(\sigma(\alpha))$ and $tr_{L:K}(\alpha) = tr_{L:K}(\sigma(\alpha))$ for each $\sigma \in G$, norm and trace are fixed by G; (i) follows from this. The remaining results follow from (i), and the definitions.   □

We also have the following transitivity result.

Proposition 11.8 *Suppose that $L : K$ is a Galois extension of degree n, with Galois group G, that H is a normal subgroup of G and that F is the fixed field of H. If $\alpha \in L$ then $N_{L:K}(\alpha) = N_{F:K}(N_{L:F}(\alpha))$ and $tr_{L:K}(\alpha) = tr_{F:K}(tr_{L:F}(\alpha))$.*

Proof   Let $H = \{\tau_1, \ldots, \tau_k\}$ and let the cosets of H in G be $\{H\rho_1, \ldots, H\rho_l\}$. Then $G = \{t_i\rho_j : 1 \leq i \leq k, 1 \leq j \leq l\}$. Thus

$$N_{L:K} = \prod_{j=1}^{l} \rho_j \left(\prod_{i=1}^{k} \tau_i(\alpha) \right) = N_{F:K}(N_{L:F}(\alpha)),$$

$$tr_{L:K} = \sum_{j=1}^{l} \rho_j \left(\sum_{i=1}^{k} \tau_i(\alpha) \right) = tr_{F:K}(tr_{L:F}(\alpha)).$$
                    □

Corollary 11.9 *Suppose that $L : K$ is a Galois extension of degree n and that $\alpha \in L$. If the minimal polynomial of α over K is $x^r + a_1 x^{r-1} + \cdots + a_r$ and $n = rs$ then $N_{L:K}(\alpha) = (-1)^n a_r^s$ and $tr_{L:K}(\alpha) = -sa_1$.*

Proof

$$N_{L:K}(\alpha) = N_{K(\alpha):K}(N_{L:K(\alpha)}(\alpha))$$
$$= N_{K(\alpha):K}(\alpha^s) = ((-1)^r a_r)^s = (-1)^n a_r^s.$$

The proof for the trace is exactly similar.   □

Theorem 11.10 (Hilbert's Theorem 90) *Suppose that $L : K$ is a Galois extension with cyclic Galois group G generated by τ and that $\alpha \in L$. Then $N_{L:K}(\alpha) = 1$ if and only if there exists $\beta \in L$ such that $\alpha = \beta/\tau(\beta)$.*

Proof Since $N_{L:K}(\beta) = N_{L:K}(\tau(\beta))$, the condition is sufficient. Suppose that $N_{L:K}(\alpha) = 1$ and that G has order n. Let $\gamma_0 = \alpha$, and inductively let $\gamma_j = \alpha\tau(\gamma_{j-1}) = \prod_{i=0}^{j-1} \tau^i(\alpha)$ for $1 \le j \le n - 1$. Since the set $\{\tau^j : 0 \le j \le n - 1\}$ is a set of distinct characters, its elements are linearly independent, and so there exists $\delta \in L^*$ such that

$$\beta = \gamma_0\delta + \gamma_1\tau(\delta) + \cdots + \gamma_{n-1}\tau^{n-1}(\delta) \ne 0.$$

Since $\tau(\gamma_{j-1}) = N_{L:K}(\alpha) = 1$, it follows that $\alpha\tau(\beta) = \beta$. ☐

Exercises

11.19 Suppose that $L : K$ is a Galois extension and that $tr_{L:K}$ is the trace. Let $T(\alpha, \beta) = tr_{L:K}(\alpha\beta)$, for $\alpha, \beta \in L$. Show that T is a K-valued bilinear form on L. If (u_i) is a basis for L over K, what is the corresponding matrix for T? Show that T is non-singular.

11.20 Suppose that $L : K$ is a Galois extension which is a splitting field extension for $f \in K[x]$, and that $\alpha_1, \ldots, \alpha_r$ are the roots of f in L. Let $M = (T(\alpha_i, \alpha_j))_{i,j=1,1}^{r,r}$, where T is the bilinear form of the previous question. Show that M is the discriminant of f.

11.7 A Sufficient Condition

We now turn to the converse of Theorem 11.10.

Theorem 11.11 *Suppose that $L : K$ is a cyclic extension of degree n, that char K does not divide n and that K contains a primitive nth root of unity, ω say. Then there exists θ in K such that $x^n - \theta$ is irreducible over K and $L : K$ is a splitting field extension for $x^n - \theta$. If β is a root of $x^n - \theta$ in L, then $L = K(\beta)$.*

Proof Since $N_{L:K}(\omega) = 1$, it follows from Hilbert's Theorem 90 that there exists $\beta \in L$ such that $\beta = \omega\beta$. Alternatively, we can argue directly.

Let σ be a generator for the cyclic group $\Gamma(L : K)$. Since the identity, $\sigma, \sigma^2, \ldots, \sigma^{n-1}$ are distinct automorphisms of L, by the corollary to Theorem 3.10 there exists α in L such that

$$\beta = \alpha + \omega\sigma(\alpha) + \cdots + \omega^{n-1}\sigma^{n-1}(\alpha) \ne 0.$$

Observe that $\sigma(\beta) = \omega^{-1}\beta$: this means first that $\beta \notin K$ and second that $\sigma(\beta^n) = (\sigma(\beta))^n = \beta^n$, so that $\theta = \beta^n \in K$.

As

$$x^n - \theta = (x - \beta)(x - \omega\beta) \ldots (x - \omega^{n-1}\beta),$$

$K(\beta) : K$ is a splitting field extension for $x^n - \theta$ over K. Since the identity, $\sigma, \ldots, \sigma^{n-1}$ are distinct automorphisms of $K(\beta)$ which fix K,

$$[K(\beta) : K] = |\Gamma(K(\beta) : K)| \geqslant n$$

and so $L = K(\beta)$. The irreducibility of $x^n - \theta$ over K now follows from Theorem 11.5.   □

Exercises

11.21 Suppose that $[L : K]$ is a prime p, that $p \neq \operatorname{char} K$ and that L is algebraically closed. Suppose (if possible) that $p > 2$.

   (i) Show that the cyclotomic polynomials Φ_p and Φ_{p^2} split over K.

   (ii) Show that there exists θ in K such that $x^p - \theta$ is irreducible over K and $L : K$ is the splitting field extension for $x^p - \theta$.

   (iii) Show that $f = x^{p^2} - \theta$ has no roots in K, and must be of the form $f = f_1 \ldots f_p$, where each f_j is an irreducible polynomial in $K[x]$ of degree p.

   (iv) Show that if $\alpha_1, \ldots, \alpha_p$ are roots of f_1 then $\alpha_1 \ldots \alpha_p = \omega\beta$, where ω is a p^2th root of unity and $\beta^p = \theta$. Explain why this gives a contradiction.

11.22 Suppose that $[L : K] = 4$, that $\operatorname{char} K \neq 2$ and that L is algebraically closed. Show that there exists an intermediate field M such that $[L : M] = 2$ and such that Φ_4 splits over M. Show that this leads to a contradiction.

11.23 Suppose that $\operatorname{char} K = 0$, that $1 < [L : K] < \infty$ and that L is algebraically closed. Show that $[L : K] = 2$ and that $L : K$ is a splitting field extension for $x^2 + 1$. (You will probably need the fact that it p is a prime which divides the order of a group G then G has a subgroup of order p.)

    The next three exercises are concerned with cyclic extensions of degree p, in the case where $\operatorname{char} K = p$.

11.24 Suppose that $\operatorname{char} K = p$, that $f = x^p - x - \alpha \in K[x]$ and that $L : K$ is a splitting field extension for f. Show that if β is a root of f then the roots of f are $\beta, \beta + 1, \ldots, \beta + p - 1$. Show that either f splits over K or f is irreducible over K and $L : K$ is cyclic of degree p.

11.25 Suppose that $L : K$ is a Galois extension with Galois group G. If $x \in L$, let

$$\text{tr}(x) = \sum_{\sigma \in G} \sigma(x).$$

Show that tr is a K-linear mapping of L onto K. The mapping tr is the *trace*. What is the effect of tr on K if char $K | G|$?

11.26 Suppose that char $K = p$, that $L : K$ is a cyclic extension of degree p and that τ generates $\Gamma(L : K)$. Let z be an element of L with tr$(z) = 1$, and let

$$y = (p-1)z + (p-2)\tau(z) + \cdots + 2\tau^{p-3}(z) + \tau^{p-2}(z).$$

Show that $\tau(y) - y = 1$, and that $\alpha = y^p - y \in K$. Show that $f = x^p - x - \alpha$ is irreducible over K, that $L : K$ is a splitting field extension for f and that $L = K(y)$.

11.27 Suppose that $L : K$ is a Galois extension of degree n with Galois group G. If $x \in L$, let

$$\text{tr}(x) = \sum_{\sigma \in G} \sigma(x), \quad N(x) = \prod_{\sigma \in G} \sigma(x).$$

The mapping N is the *norm*. Suppose that $\alpha \in L$ has minimal polynomial

$$x^r - a_1 x^{r-1} + \cdots + (-1)^r a_r$$

Show that $\text{tr}(\alpha) = (n/r)a_1$ and $N(\alpha) = a_r^{n/r}$.

11.8 Kummer Extensions

It is not difficult to extend Theorems 11.5 and 11.11 to Galois extensions with abelian Galois groups.

Theorem 11.12 *Suppose that $L : K$ is a Galois extension, that $\Gamma(L : K)$ is an abelian group of exponent d and that $x^d - 1$ has d distinct roots in K. Then there exist $\theta_1, \ldots, \theta_r$ in K such that $L : K$ is a splitting field extension for*

$$(x^d - \theta_1) \ldots (x^d - \theta_r).$$

Proof The proof is by induction on $[L : K]$. Suppose that $[L : K] = n$ and that the result holds for all extensions of smaller degree.

By Theorem 1.8 we can write

$$\Gamma(L : K) = F \times H$$

where F is cyclic of order d and H is an abelian group whose exponent e divides d. Let M be the fixed field of F; then $M : K$ is a Galois extension, and $\Gamma(M : K) \cong H$. As $x^e - 1$ has e distinct roots in K and $[M : K] < n$, by the inductive hypothesis there exist $\psi_1, \ldots, \psi_{r-1}$ in K such that $M : K$ is a splitting field extension for

$$(x^e - \psi_1) \ldots (x^e - \psi_{r-1}).$$

Let β_j be a root of $x^e - \psi_j$ in M, let $\theta_j = \psi_j^{d/e}$ and let ω be a primitive dth root of unity in K. Then

$$x^d - \theta_j = \prod_{i=0}^{d-1} (x - \omega^i \beta_j)$$

so that $M : K$ is a splitting field extension for

$$(x^d - \theta_1) \ldots (x^d - \theta_{r-1}).$$

We now argue as in Theorem 11.11. Let σ be a generator for F. As $\Gamma(L : K) = \{\sigma^j \tau : 0 \leqslant j < d, \tau \in H\}$, there exists α in L such that

$$\beta_r = \sum_{\tau \in H} \tau(\alpha) + \omega\sigma \sum_{\tau \in H} \tau(\alpha) + \cdots + \omega^{d-1}\sigma^{d-1} \sum_{\tau \in H} \tau(\alpha) \neq 0.$$

As before, $\sigma(\beta_r) = \omega^{-1}\beta_r$, so that $\sigma(\beta_r^d) = (\sigma(\beta_r))^d = \beta_r^d$ and $\theta_r = \beta_r^d \in M$. But also $\tau(\beta_r) = \beta_r$ for $\tau \in H$, so that $\tau(\theta_r) = \theta_r$ for $\tau \in H$, and so $\theta_r \in K$. As in Theorem 11.11, $L : M$ is a splitting field extension for $x^d - \theta_r$, and so $L : K$ is a splitting field extension for $(x^d - \theta_1) \ldots (x^d - \theta_r)$.   □

As extension $L : K$ is called a *Kummer extension of exponent d* if it is a splitting field extension of a polynomial of the form

$$(x^d - \theta_1) \ldots (x^d - \theta_r)$$

(where $\theta_1, \ldots, \theta_r$, are in K) and if $x^d - 1$ has d distinct roots in K.

Let us now prove a converse to Theorem 11.12.

Theorem 11.13 *Suppose that $L : K$ is a Kummer extension of exponent d. Then $\Gamma(L : K)$ is abelian, and its exponent divides d.*

Proof Suppose that $L : K$ is a splitting field extension of

$$f = (x^d - \theta_1) \ldots (x^d - \theta_r).$$

By Theorem 9.6, we need only consider the action of $\Gamma(L : K)$ on the roots of f. Let ω be a primitive dth root of unity in K. Then if $\sigma \in \Gamma(L : K)$ and β_j is a root of $x^d - \theta_j$ in L, $\sigma(\beta_j) = \omega^{n_{\sigma,j}}\beta_j$ for some $n_{\sigma,j}$, so that $\sigma^d(\beta_j) = \beta_j$, and $\sigma^d = e$. This implies that the exponent of $\Gamma(L : K)$ divides d. If τ is another element of $\Gamma(L : K)$,

$$\tau\sigma(\beta_j) = \tau(\omega^{n_{\sigma,j}}\beta_j) = \omega^{n_{\sigma,j}}\omega^{n_{\tau,j}}\beta_j$$
$$= \sigma(\omega^{n_{\tau,j}}\beta_j) = \sigma\tau(\beta_j),$$

so that $\Gamma(L : K)$ is abelian. $\square$

Exercises

11.28 Suppose that K is a field which contains a primitive nth root of unity and that $x^n - a$ and $x^n - b$ are irreducible over K. Show that if $b = a^r c^n$ for some r which is prime to n and some c in K, then $x^n - a$ and $x^n - b$ have the same splitting field extension over K.

11.29 Suppose K is a field which contains a primitive nth root of unity, and that $x^n - a$ and $x^n - b$ are irreducible polynomials over K with the same splitting field extension $L : K$. Let α be a root of $x^n - a$ in L, β a root of $x^n - b$. By considering the action of $\Gamma(L : K)$ on α and β, show that there exists r, prime to n, such that $\beta\alpha^{-r} \in K$. Show that $b = a^r c^n$ for some c in K.

12

Solution by Radicals

The results of the two preceding chapters, together with the fundamental theorem of Galois theory, suggest that, provided that we can construct enough roots of unity, a separable polynomial is solvable by radicals if and only if its Galois group can be built up in some way from abelian groups. We shall see that this is indeed so.

12.1 Polynomials with Soluble Galois Groups

In this section we shall show that, if f is separable and has a soluble Galois group, then, provided that we can construct enough roots of unity, f is solvable by radicals.

Theorem 12.1 *Suppose that f is a separable polynomial in $K[x]$ whose Galois group $\Gamma_K(f)$ is soluble, and suppose that* char K *does not divide* $|\Gamma_K(f)|$. *Then f is solvable by radicals.*

Proof The proof is largely a matter of putting together results that we have already established. Let $d = |\Gamma_K(f)|$. If K does not contain a primitive dth root of unity, we can adjoin one, ε say. Let $L = K(\varepsilon)$. $L : K$ is an extension by radicals. Since char K does not divide d, L contains d distinct dth roots of unity. Now let $N : L$ be a splitting field extension for f over K. $N : L$ is a Galois extension, and by the theorem on natural irrationalities (Theorem 17.1) $\Gamma(N : L) = \Gamma_L(f)$ is isomorphic to a subgroup of $\Gamma_K(f)$. Thus $\Gamma_L(f)$ is soluble, by Theorem 1.24(i). This means that there exist subgroups

$$\{e\} = G_r \lhd G_{r-1} \lhd \cdots \lhd G_0 = \Gamma_L(f)$$

such that G_{j-1}/G_j is cyclic, for $1 \leqslant j \leqslant r$.

We now exploit the fundamental theorem of Galois theory. Let L_j be the fixed field of G_j. Then

$$N = L_r : L_{r-1} : \ldots : L_0 = L.$$

Also $\Gamma(N : L_{j-1}) = G_{j-1}$, and $G_j \lhd G_{j-1}$, so that, by the fundamental theorem,

$$\Gamma(L_j : L_{j-1}) \cong G_{j-1}/G_j, \text{ for } 1 \leqslant j \leqslant r.$$

Thus $L_j : L_{j-1}$ is a cyclic extension. Also $[L_j : L_{j-1}] = |G_{j-1}/G_j|$, so that $[L_j : L_{j-1}]$ divides d: thus char K does not divide $[L_j : L_{j-1}]$, and L_{j-1} contains a primitive $[L_j : L_{j-1}]$th root of unity. By Theorem 11.11, there exists an element β_j in L_j such that $L_j = L_{j-1}(\beta_j)$ and such that β_j is a radical over L_{j-1}. Thus $N : L$ is an extension by radicals, and so $N : K$ is also an extension by radicals. Since f splits over N, f is solvable by radicals.

Notice that if $f \in K[x]$ and if either char $K = 0$ or char $K > $ degree f, then f must be separable, by Theorem 8.6, and char K cannot divide $|\Gamma_K(f)|$, since $|\Gamma_K(f)|$ divides (degree f)!. $\qquad\square$

12.2 Polynomials which are Soluble by Radicals

We now turn to results in the opposite direction. Here, the main problem is one of normality. Suppose that

$$L = L_r : L_{r-1} : \ldots : L_0 = K$$

is an extension by radicals. Even if K contains sufficient many roots of unity, so that each of the extensions $L_j : L_{j-1}$ is normal, it does not follow that $L : K$ is a normal extension. We get round this difficulty by a symmetrization argument.

Theorem 12.2 *Suppose that $L : K$ is a Galois extension, that $M = L(\beta)$, where β is a root of $x^n - \theta$ (with θ in L) and that char K does not divide n. Then there exists an extension by radicals $N : M$ such that $N \cdot K$ is a Galois extension.*

Proof Since char K does not divide n we can if necessary adjoin a primitive nth root of unity, ε say, to M. Then in $M(\varepsilon)[x]$

$$x^n - \theta = (x - \beta)(x - \varepsilon\beta)\ldots(x - \varepsilon^{n-1}\beta)$$

so that $M(\varepsilon) : L$ is a splitting field extension for $x^n - \theta$ over L. As $x^n - \theta$ has n distinct roots, $M(\varepsilon) : L$ is a Galois extension. Note also that $M(\varepsilon) : L = L(\beta, \varepsilon) : L$ is an extension by radicals.

Now let $G = \Gamma(L : K)$ and let

$$f = \prod_{\sigma \in G} (x^n - \sigma(\theta)).$$

Let $N : M(\varepsilon)$ be a splitting field for f over $M(\varepsilon)$. We have the following tower of extensions:

$$N : M(\varepsilon) : M : L : K.$$

If β_σ is a root of $x^n - \sigma(\theta)$ in N then

$$x^n - \sigma(\theta) = (x - \beta_\sigma)(x - \varepsilon\beta_\sigma)\dots(x - \varepsilon^{n-1}\beta_\sigma);$$

thus $N : L$ is a splitting field extension for f over L. Also, $x^n - \sigma(\theta)$ has n distinct roots in N, for each σ, so that f is separable over $M(\varepsilon)$. As $M(\varepsilon) : L$ and $L : K$ are both separable, this means that $N : K$ is separable, by Corollary 4 to Theorem 8.3.

We now use the symmetry of f; if $\tau \in G$, $\tau(f) = f$ so that, since $L : K$ is a Galois extension, $f \in K[x]$. There exists g in $K[x]$ such that $L : K$ is a splitting field extension for g over K. Thus $N : K$ is a splitting field extension for fg over K, and so $N : K$ is normal.

Finally observe that N is obtained from M by first adjoining ε and then adjoining the roots of $x^n - \sigma(\theta)$, for σ in G, and so $N : M$ is an extension by radicals.   □

We now apply this to extensions by radicals.

Theorem 12.3 *Suppose that*

$$L = L_r : L_{r-1} : \dots : L_0 = K$$

is an extension by radicals, with $L_i = L_{i-1}(\beta_i)$, *where* β_i *is a root of* $x^{n_i} - \theta_i$ *(with* $\theta_i \in L_{i-1}$*). Then if char K does not divide* $n_1 n_2 \dots n_r$, *there exists an extension* $M : L$ *such that* $M : K$ *is a Galois extension by radicals.*

Proof We prove this by induction on r. The result is trivially true when $r = 0$. Suppose that the result holds for $r - 1$. Then there exists an extension $M_{r-1} : L_{r-1}$ such that $M_{r-1} : K$ is a Galois extension by radicals.

Let m_r be the minimal polynomial for β_r over L_{r-1}, and let l_r be an irreducible factor of m_r, considered as an element of $M_{r-1}[x]$. By Theorem 6.2, there is a simple algebraic extension $M_{r-1}(\gamma) : M_{r-1}$ such that $l_r(\gamma) = 0$.

Since this means that $m_r(\gamma) = 0$, it follows from Theorem 6.4 that there is a monomorphism i from $L = L_{r-1}(\beta_r)$ into $M_{r-1}(\gamma)$, fixing L_{r-1}, such that $i(\beta_r) = \gamma$. In other words, identifying L with $i(L)$, we can suppose that L and M_{r-1} are both subfields of $M_{r-1}(\beta_r)$.

We apply Theorem 12.2 to the Galois extension $M_{r-1} : K$ and $M_{r-1}(\beta_r)$, and conclude that there is an extension $M_r : M_{r-1}(\beta_r)$ by radicals such that $M_r : K$ is a Galois extension. We have the following diagram:

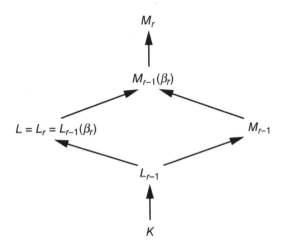

Since each of the extensions $M_r : M_{r-1}(\beta_r)$, $M_{r-1}(\beta_r) : M_{r-1}$ and $M_{r-1} : K$ is an extension by radicals, so is $M_r : K$.

Notice that the conditions on char K are also satisfied by the extension $M : K$.   □

Theorem 12.4 *Suppose that*

$$L = L_r : L_{r-1} : \ldots : L_0 = K$$

is an extension by radicals, with $L_i = L_{i-1}(\beta_i)$, where β_i is a root of $x^{n_i} - \theta_i$ (with θ_i in L_{i-1}), and that char K *does not divide $n_1 \ldots n_r$. If $f \in K[x]$ splits over L, then the Galois group $\Gamma_K(f)$ is soluble.*

Proof By Theorem 12.3 and the remark following it, we can assume that $L : K$ is a Galois extension.

For each $1 \leqslant i \leqslant r$, $L : L_i$ is a Galois extension (Corollary 3 to Theorem 7.1, and Theorem 8.1): $x^{n_i} - \theta_i$ has a root β_i in L, and so it splits over L. This means, by Theorem 11.5, that L contains a primitive n_ith root of unity. Let n be the lowest common multiple of $n_1, \ldots, n_r$: then L contains a primitive nth root of unity, ε say.

Now let $L'_i = L_i(\varepsilon)$, for $0 \leqslant i \leqslant r$. Then we have the following tower of extensions:

$$L = L'_r : L'_{r-1} : \ldots : L'_0 = L_0(\varepsilon) : L_0 = K.$$

We shall show that $G = \Gamma(L : K)$ is soluble. Let $G_i = \Gamma(L : L'_i)$, for $0 \leqslant i \leqslant r$. As $L'_i : L'_{i-1}$ is a splitting field extension for $x^{n_i} - \theta_i$ over $L'_{i-1}, L'_i : L'_{i-1}$ is cyclic, and so by the fundamental theorem of Galois theory $G_i \lhd G_{i-1}$ and $G_{i-1}/G_i \cong \Gamma(L'_i : L'_{i-1})$. Thus $G_0 = \Gamma(L : L'_0) = \Gamma(L : K(\varepsilon))$ is soluble. Now ε is a primitive nth root of unity, so that $K(\varepsilon) : K$ is a splitting field extension for $x^n - 1$, and $\Gamma(K(\varepsilon) : K)$ is abelian (Theorem 11.4), and therefore soluble. By the fundamental theorem of Galois theory,

$$\Gamma(K(\varepsilon) : K) \cong \Gamma(L : K)/\Gamma(L : K(\varepsilon)) = G/G_0$$

so that G/G_0 is soluble. Consequently G is soluble, by Theorem 1.24.

Remember that f splits over L. Let $N : K$ be a splitting field extension for f over K, with $N \subseteq L$. The extension $N : K$ is normal; using the fundamental theorem of Galois theory once again,

$$\Gamma_K(f) = \Gamma(N : K) \cong \Gamma(L : K)/\Gamma(L : N) = G/\Gamma(L : N),$$

and so $\Gamma_K(f)$ is soluble, by Theorem 1.24. $\square$

As an example, we have seen that the quintic $x^5 - 4x + 2$ is irreducible over $\mathbb{Q}$, and has Galois group Σ_5. Σ_5 is not soluble, and so $x^5 - 4x + 2$ cannot be solved by radicals!

Exercises

12.1 Let $L_0 = \mathbb{Q}$, $L_1 = \mathbb{Q}(3^{1/2})$, $L_2 = \mathbb{Q}((3^{1/2} + 1)^{1/2})$. Show that $L_1 : L_0$ and $L_2 : L_1$ are both normal extensions but that $L_2 : L_0$ is not normal. Find the minimal polynomial of $(3^{1/2} + 1)^{1/2}$ over $\mathbb{Q}$, and find its Galois group.

12.2 Let f be an irreducible cubic in $K[x]$, where K is a subfield of $\mathbb{R}$. Show that f has three real roots if and only if its discriminant is positive.

12.3 Suppose that K is a subfield of $\mathbb{R}$ and that f is an irreducible cubic in $K[x]$ with three real roots. Suppose that $L = K(r)$, where $r \in \mathbb{R}$ and $r^p \in K$ for some prime p. Show that f is irreducible over L.

12.4 Suppose that K is a subfield of $\mathbb{R}$ and that f is an irreducible cubic in $K[x]$ with three real roots. Show that if $L : K$ is an extension by radicals with $L \subseteq \mathbb{R}$ then f is irreducible over L. (It is not possible to solve f only by extracting real roots!)

12.5 Give an example of a polynomial in $\mathbb{Q}[x]$ which is solvable by radicals, but whose splitting field is not an extension by radicals.

12.6 Suppose that $L : K$ is an extension by radicals and that M is an intermediate field. Is $M : K$ an extension by radicals?

13

Regular Polygons

In this chapter, we consider which regular polygons can be constructed by ruler and compasses. This requires more than the material of Chapter 5. Indeed, the chapter can be seen as an extended exercise of the theory of Chapters 5 to 8, together with the theory of p-groups and soluble groups. We need one more ingredient.

13.1 Fermat Primes and Fermat Numbers

A Fermat prime is a prime number of the form $2^k + 1$.

Proposition 13.1 *A Fermat prime is of the form $2^{2^r} + 1$, for some integer r.*

Proof For if $u = 2^{rs}$, where s is odd, then

$$u + 1 = (2^r + 1)(2^{r(s-1)} - 2^{r(s-2)} + \cdots - 2^r + 1),$$

so that $u + 1$ is not prime. □

This leads us to define the *nth Fermat number* to be $F_n = 2^{2^n} + 1$. F_n is a *Fermat prime* if it is also prime. Then F_n is a Fermat prime for $0 \le n \le 4$; $F_0 = 3, F_1 = 5, F_2 = 17, F_3 = 257$ and $F_4 = 65,537$. No other Fermat primes are known, and indeed it is not known if there are infinitely many Fermat primes. Many Fermat numbers are known to be composite (not prime); they provide a useful subject for computer factorization programs.

Exercise

13.1 Suppose that $n = bc$, where b is odd. Show that $a^n + 1$ is divisible by $a^c + 1$. Deduce that if $a^k + 1$ is a prime number, then a is even and $k = 2^j$ for some j.

13.2 Regular Polygons

Let us now set the scene. We consider a k-sided regular polygon Π_k in the complex plane with centre the origin, and with a vertex at 1. Thus if $a_k = e^{2\pi i/k}$ then the elements of the cyclic group G_k generated by a_k are the vertices of Π_k. We say that Π_k is *constructible* if its vertices can be constructed by ruler and compasses. Here are some simple results about constructible polygons.

Proposition 13.2 Π_k *is constructible if and only if* Π_{2k} *is constructible, and if k and l have no common divisor then Π_{kl} is constructible if and only if Π_k and Π_l are both constructible.*

Proof Since the bisector of angles is constructible by ruler and compasses, Π_{2k} is constructible if Π_k is constructible, and l divides k, then $G_{l}l$ is a subgroup of G_k, so that Pi_l is constructible. If k and l have no common factor then $G_{kl} = Gp(G_k \cup G_l)$, so that Π_{kl} is constructible if and only if Π_k and Π_l are. $\qquad\square$

We are now in a position to determine when a regular polygon is constructible.

Theorem 13.3 (Gauss' theorem) *Suppose that* $k = 2^n q_1^{l_1} \ldots q_r^{l_r}$, *where* $q_1, \ldots, q_r$ *are distinct prime numbers. Then the regular polygon Π_k is constructible if and only if $q_1, \ldots, q_r$ are distinct Fermat primes, and* $l_1 = \cdots = l_r = 1$.

Proof We begin with necessity. By the preceding proposition, it is sufficient to show that if Π_k is constructible, where $k = q^l$ (with q a prime) then q is a Fermat prime and $l = 1$. If $l > 1$, then Π_{q^2} is constructible. Let $m = q^2$, and let e_m be a primitive mth root of unity. But then by Theorem 11.4, $\Gamma_{\mathbb{Q}}(\Phi_m) \simeq U_m$, so that $|\Gamma_{\mathbb{Q}}(\Phi_m)| = q(q-1)$, and Π_m is not constructible. Thus $l = 1$, and so $|\Gamma_{\mathbb{Q}}(\Phi_q)| = q - 1$, and q is a Fermat prime.

Conversely suppose that $n = 2^s + 1$ is prime. Then $[\mathbb{Q}(\varepsilon_n) : \mathbb{Q}] = 2^s$, $\mathbb{Q}(\varepsilon_n) : \mathbb{Q}$ is a splitting field extension for Φ_n, and $G = \Gamma(\mathbb{Q}(\varepsilon_n) : \mathbb{Q})$ is cyclic of degree 2^s, by the corollary to Theorem 11.4. Let σ be a generator for G. Then, if $0 \leqslant t \leqslant s$, the group G_{s-t} generated by $\sigma^{2^{s-t}}$ has order 2^t, and so there are intermediate groups

$$\{e\} = G_s \subseteq G_{s-1} \subseteq \cdots \subseteq G_0 = G, \text{ with } |G_{j-1}/G_j| = 2, \text{ for } 1 \leqslant j \leqslant s.$$

Let L_j be the fixed field for G_j. Then

$$\mathbb{Q}(\varepsilon_n) = L_s : L_{s-1} : \ldots : L_0 = \mathbb{Q}$$

is a tower of fields, and $[L_j : L_{j-1}] = 2$ for $1 \leqslant j \leqslant s$.

We shall show that if $z \in \mathbb{Q}(\varepsilon_n)$ and $z = x + iy$ then (x, y) is constructible. We use induction on j. The result is true for elements of $L_0 = \mathbb{Q}$. Suppose that it holds for all elements of L_{j-1} and that $\alpha = \alpha_1 + i\alpha_2 \in L_j \backslash L_{j-1}$. Then the minimal polynomial m_α for α over L_{j-1} is a quadratic in $L_{j-1}[x]$:

$$m_\alpha = x^2 + 2bx + c.$$

Thus $\alpha = -b + \nu$, where $\nu^2 = \mu = b^2 - c \in L_{j-1}$. Let us set $b = b_1 + ib_2$, $\nu = \nu_1 + i\nu_2$ and $\mu = \mu_1 + i\mu_2 = re^{i\theta}$. By the inductive hypothesis, (μ_1, μ_2) is constructible: from this we can successively construct $(r, 0)$, $(r^{1/2}, 0)$ and $(r^{1/2} \cos(\theta/2), r^{1/2} \sin(\theta/2))$ (since we can bisect angles). As $(\nu_1, \nu_2) = \pm(r^{1/2} \cos(\theta/2), r^{1/2} \sin(\theta/2))$ and as (b_1, b_2) is constructible, by the inductive hypothesis, this means that (α_1, α_2) is constructible. This establishes the induction. This means that (x_n, y_n) is constructible (where $\varepsilon_n = x_n + iy_n$) and so n is possible.   □

13.3 Constructing a Regular Pentagon

As an illustration, let us consider the construction of a regular pentagon. Although ingenious constructions exist, we shall consider a straightforward one which highlights the algebra. Let $\alpha = e^{2\pi i/5}$ and let $\beta = \alpha + \alpha^4 = 2\cos(2\pi/5)$. Then $\beta^2 + \beta - 1 = 0$, so that $\cos(2\pi/5) = (-1 + \sqrt{5})/2$, which is a constructible point on the real line. Construct the perpendicular to the real line through $\beta/2$: this meets the unit circle at α and α^4; α^2 and α^3 are constructed in the same way.

Exercises

13.2 Let $f(x) = x^4 + x^3 + x^2 + x + 1 \in \mathbb{Z}[x]$. Identify its splitting field L as a subfield of $\mathbb{Q}$. Identify its Galois group as a group of automorphisms of L. What are the intermediate fields?

13.3 Find a quintic in $f \in \mathbb{Z}[x]$ such that $f(\cos(\theta)) = \cos(5\theta)$. Can $x = \cos(\pi/10)$ be constructed by radicals? What about $x = \cos(\pi/15)$?

13.4 Find a cubic in $\mathbb{Z}[x]$ which has $\cos(2\pi/7)$ as a root, and a quartic in $\mathbb{Z}[x]$ which has $\cos(2\pi/9)$ as a root. Are these numbers constructible by radicals?

14

Polynomials of Low Degree

Although we shall obtain further general results in Galois theory in the next chapter, Galois theory's first intent is to throw light on the solution of polynomial equations. We now pause to see how the theory that we have developed so far relates to the solution of equations of low degree. In this chapter, we consider the solution of monic quadratic, cubic and quartic polynomial equations over a field K. Solution of these equations shows how Galois theory is used, and also suggests how the theory will develop further.

14.1 Quadratic Polynomials

First we consider irreducible quadratic polynomials in $\mathbb{Q}[x]$. First consider $f(x) = x^2 + 1$. This takes positive values on $\mathbb{Q}$, and so is irreducible. (Also its discriminant Δ is -1, which does not have a square root in $\mathbb{Q}$.) We extend by adding an element i for which $i^2 + 1 = 0$, thus obtaining the field $\mathbb{Q}(i)$, over which $x^2 + 1 = (x - i)(x + i)$; i and $-i$ are the roots of f in $\mathbb{Q}(i)$. Any element of $\mathbb{Q}(i)$ is of the form $a + ib$, with $a, b \in \mathbb{Q}$; we are led to consider complex numbers. $[\mathbb{Q}(i) : \mathbb{Q}] = 2$, and the Galois group is Σ_2, with two elements, the identity, and the transposition of i and $-i$.

If $f(x) = x^2 + ax + b$ is a monic quadratic polynomial in $\mathbb{Q}[x]$, we proceed by '*completing the square*'. Let $y = x + a/2$, so that $f(x) = y^2 - \mu$, where $\mu = (a^2 - 4b)/4 = \Delta/4$. We therefore consider the polynomial $g(x) = x^2 - \mu$, and if necessary add an element $\sqrt{\mu}$. Then the roots of f are $a/2 + \sqrt{\mu}$ and $a/2 - \sqrt{\mu}$. If μ does not have a square root in $\mathbb{Q}$, then f has $\mathbb{Q}(\sqrt{\mu})$ as splitting field, and again the Galois group has two elements, the identity, and the transposition of the two roots.

As an example, let $f(x) = x^2 + x + 1 \in \mathbb{Q}[x]$. If $y = x + 1/2$, then $f(x) = y^2 + 3/4$, so that the splitting field is $\mathbb{Q}(\sqrt{-3/4}) = \mathbb{Q}(\sqrt{-3})$, and the roots are $\omega = \frac{1}{2}(-1+\sqrt{-3})$ and $\omega^2 = \frac{1}{2}(-1-\sqrt{-3})$. Although it is tempting to write $i\sqrt{3}$ for $\sqrt{-3}$, neither i nor $\sqrt{3}$ lie in the splitting field. Notice also that $\omega^3 = 1$; we shall consider the polynomial x^2+x+1 further in the next section.

What happens if we replace $\mathbb{Q}$ by a more general field K? There is no difficulty if the characteristic of K is 0, but care is needed if $(K) = p > 0$. For example, if $(K) = 2$, then $x^2 + 1 = (x - 1)^2$, so that $x^2 + 1$ has a repeated root in $\mathbb{Z}_2$. But $g(x) = X^2 + x + 1$ is irreducible in $\mathbb{Z}_2$.

Exercise

14.1 Show that $f(x) = x^2+x+1$ has a repeated root in $\mathbb{Z}_3$ and is irreducible in $\mathbb{Z}_5$. What happens in $\mathbb{Z}_7$? For what values of p less than 40 is f irreducible?

In the same way, we can also consider quadratic polynomials $f(x) = x^2 - a$ in $K(x)$, where a is not a square in K, to obtain an extension $K(\sqrt{a})$, with similar properties.

Exercise

14.2 If a and b are not squares in K and $a \neq b$, then $K(\sqrt{a})$ is isomorphic to $K(\sqrt{b})$ if and only if a/b is a square in K.

Clearly if f is a separable irreducible quadratic polynomial in $K(x)$ then its Galois group is isomorphic to $\mathbb{Z}_2$.

14.2 Cubic Polynomials

We consider cubic polynomials over a field K, where the characteristic of K is not 2 or 3. First we consider $f(x) = x^3 - 1$. This is not irreducible, but factorizes as $f(x) = (x-1)(x^2+x+1)$. (If char $(K) = 3$, then $f(x) = (x^2-1)$ $(x + 1) = (x - 1)^2(x + 1)$, which is why we suppose that the characteristic of K is not 3.) As we have seen, if $x^2 + x + 1$ is irreducible over K we adjoin a root ω of $g(x) = x^2 + x + 1$, then $x^3 = (x - 1)(x - \omega)(x - \omega^2)$; $1, \omega$ and ω^2 are the three cube roots of 1, and $[K(\omega) : K] = 2$. Similarly, $-1, -\omega$ and $-\omega^2$ are the three cube roots of -1.

Suppose now that $f(x) = x^3 + ax^2 + bx + c$ is a monic irreducible cubic polynomial in $K(x)$. As with quadratic polynomials, we simplify things. Let $y = x + a/3$. Then $f(x) = y^3 + py + q$, where $p = b - a^2/3$ and $q = c + 2a^2/27 - ab/3$, and so we consider the irreducible polynomial

$f(x) = x^3 + px + q$. Let $L : K$ be a splitting field extension for f, and let $\alpha_1, \alpha_2, \alpha_3$ be the roots of f in L (which we are trying to calculate); thus $L = K(\alpha_1, \alpha_2, \alpha_3)$. Then $\Gamma_K(f) = A_3$ or Σ_3 and $[L : K] = 3$ or 6.

How do we discriminate between the two cases? Let Δ be the discriminant of f. Then $\Delta \in K$. If Δ has a square root δ in K, then $[L : K] = 3$, and $\Gamma_K(f) = A_3$, which is isomorphic to Z_3. Otherwise we adjoin a square root δ. Since $\delta = (\alpha_1 - \alpha_2)(\alpha_1 - \alpha_3)(\alpha_2 - \alpha_3)$, $\delta \in L$, $K(\delta) : K$ is a splitting field extension for $x^2 - \Delta$ and $[K(\delta) : K] = 2$, while $L : K(\delta)$ is a splitting field extension for f (considered as an element of $K(\delta)[x]$), with Galois group A_3.

Exercise

14.3 Give an example of an irreducible polynomial in $\mathbb{Z}[x]$ whose Galois group is A_3, and one for which the Galois group is Σ_3.

We now make the fundamental construction. First, if $x^3 - 1$ does not split over $K(\delta)$, we adjoin a root ω of $x^2 + x + 1$ to $K(\delta)$. If $\omega \notin K(\delta)$ then $\omega \notin L$, since $[L : K(\delta)] = 3$ and $[K(\delta, \omega) : K(\delta)] = 2$.

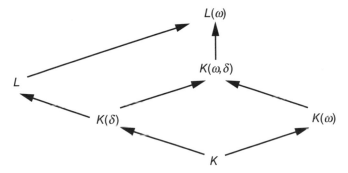

Next, we consider the elements

$$\beta = \alpha_1 + \omega \alpha_2 + \omega^2 \alpha_3 \quad \text{and} \quad \gamma = \alpha_1 + \omega^2 \alpha_2 + \omega \alpha_3$$

in $L(\omega)$. Then

$$\beta\gamma = \alpha_1^2 + \alpha_2^2 + \alpha_3^2 + (\omega + \omega^2)(\alpha_1\alpha_2 + \alpha_1\alpha_3 + \alpha_2\alpha_3)$$
$$= (\alpha_1 + \alpha_2 + \alpha_3^2) = -3p,$$

so that $\beta^3 \gamma^3 = -27p^3$ and

$$\beta^3 + \gamma^3 = (\alpha_1 + \alpha_2 + \alpha_3)^3 + (\alpha_1 + \omega\alpha_2 + \omega^2\alpha_3)^3 + (\alpha_1 + \omega^2\alpha_2 + \omega\alpha_3)^3$$
$$= 3(\alpha_1^3 + \alpha_2^3 + \alpha_3^3) + 18\alpha_1\alpha_2\alpha_3$$
$$- 3((\alpha_1 + \alpha_2 + \alpha_3) + 3q) - 18q = -27q,$$

so that β^3 and γ^3 are roots of the quadratic equation $x^2 + 27qx - 27p^3 = 0$ in $K(x)$, and so β^3 and q^3 are the elements

$$\tfrac{1}{2}(-27q + 3(2\omega + 1)\delta) \quad \text{and} \quad \tfrac{1}{2}(-27q - 3(2\omega + 1)\delta)$$

in $K(\delta, \omega)$. We can now obtain β by adjoining a cube root; then $\gamma = -3p/\beta$. Finally,

$$\alpha_1 = (\beta + \gamma)/3, \quad \alpha_2 = (\omega^2\beta + \omega\gamma)/3, \quad \alpha_3 = (\omega\beta + \omega^2\gamma)/3.$$

The procedure for solving cubic equations was developed (not in these terms!) by Tartaglia in the 1530s, and appeared in Cardan's *Ars Magna* in 1545; it is known as *Cardan's formula*.

Theorem 14.1 (Cardan's formula) *Suppose that $f(x) = x^3 + a_2x^2 + a_1x + a_3$ is a monic cubic polynomial in $K(x)$, where $K \neq 3$. Let*

$$p = a_1 - a_2^2/3, \qquad q = q_0 + 2a_2^3/27 - a_1/3,$$
$$\Delta = 4p^2 - 27q^3, \qquad \delta = \sqrt{\Delta},$$
$$b = \tfrac{1}{2}(27q + 3\sqrt{3}\delta i), \quad c = \tfrac{1}{2}(27q - 3\sqrt{3}\delta i),$$
$$\beta = \sqrt[3]{b}, \qquad \gamma = -3p/\beta.$$

Then the three roots of f in a splitting field L are

$$(\beta + \gamma - a_2)/3, \quad (\omega^2\beta + \omega\gamma - a_2)/3, \quad (\omega\beta + \omega^2\gamma - a_2)/3,$$

where $\omega = (-1 + \sqrt{3}i)/2$ is a complex cube root of 1.

Notice that again, to solve a cubic equation, we only have to adjoin square roots and cube roots.

Exercise

14.4 Suppose that char K is not 2 or 3, and that $f = x^3 + px + q \in K[x]$. Let α be a root of f in a splitting field, let $g = 3x^2 - 3\alpha x - p$, and let β be a root of g in a splitting field for g over $K(\alpha)$. Express α in terms of β and show that β is a root of

$$h = 27x^6 + 27q^3 - p^3 \in K[x].$$

Conclude that $\alpha = \beta - p/3\beta$, where $\beta^3 = -q/2 + \delta$ and $\delta^2 = q^2/4 + p^3/27$: the cubic f can be solved by extracting a square root and a cube root.

14.3 Quartic Polynomials

Suppose now that f is an irreducible monic quartic in $K[x]$:

$$f = x^4 + a_3 x^3 + a_2 x^2 + a_1 x + a_0.$$

We continue to suppose that char K is not equal to 2 or 3. If we write $y = x + a_3/4$, f has the form

$$g = y^4 + py^2 + qy + r.$$

We therefore consider a polynomial of the form

$$g = x^4 + px^2 + qx + r.$$

Let $L : K$ be a splitting field extension for g over K, and let $\alpha_1, \alpha_2, \alpha_3$ and α_4 be the roots of g in L.

Let G be the Galois group $\Gamma_K(g)$. G can be considered, by its action on the roots of g, as a transitive subgroup of Σ_4. Now the Vierergruppe N is a normal subgroup of Σ_4, and so $H = N \cap G$ is a normal subgroup of G. Let M be the fixed field of H. Then by the fundamental theorem of Galois theory $\Gamma(L : M) = H$ and $\Gamma(M : K) \cong G/H$.

Now H is an abelian group of order 1, 2 or 4 (in fact the first possibility cannot arise) and, as H is the kernel of the homomorphism ϕi, where i is the inclusion map $G \to \Sigma_4$ and ϕ is the epimorphism of Σ_4 onto Σ_3 described in Chapter 1, G/H is isomorphic to a subgroup of Σ_3, by the first isomorphism theorem for groups.

This suggests that we should first attempt to determine the intermediate field M. Let

$$\beta = \alpha_1 + \alpha_2, \quad \gamma = \alpha_1 + \alpha_3 \quad \text{and} \quad \delta = \alpha_1 + \alpha_4.$$

Then

$$\beta^2 = (\alpha_1 + \alpha_2)^2 = -(\alpha_1 + \alpha_2)(\alpha_3 + \alpha_4),$$

$$\gamma^2 = (\alpha_1 + \alpha_3)^2 = -(\alpha_1 + \alpha_3)(\alpha_2 + \alpha_4),$$

and

$$\delta^2 = (\alpha_1 + \alpha_4)^2 = -(\alpha_1 + \alpha_4)(\alpha_2 + \alpha_3).$$

Consequently β^2, γ^2 and δ^2 are in M, and so $K(\beta^2, \gamma^2, \delta^2) \subseteq M$. On the contrary, if σ is a permutation of $\alpha_1, \alpha_2, \alpha_3$ and α_4 which fixes β^2, γ^2 and δ^2, then $\sigma \in N$. Thus

$$\Gamma(L : K(\beta^2, \gamma^2, \delta^2)) \subseteq H = \Gamma(L : M)$$

and so $K(\beta^2, \gamma^2, \delta^2) \supseteq M$. Thus $M = K(\beta^2, \gamma^2, \delta^2)$.

Easy but tedious calculations show that

$$\beta^2 + \gamma^2 + \delta^2 = -2p,$$
$$\beta^2\gamma^2 + \beta^2\delta^2 + \gamma^2\delta^2 = p^2 - 4r,$$

and

$$\beta\gamma\delta = -q;$$

thus $K(\beta^2, \gamma^2, \delta^2) : K$ is a splitting field extension for

$$x^3 + 2px^2 + (p^2 - 4r)x - q^2.$$

This cubic is called the *cubic resolvent* for g. By the results of the previous section, we can construct β^2, γ^2 and δ^2 by adjoining square roots and cube roots; we can then construct β, γ and δ by adjoining square roots (note, though, that $\beta\gamma\delta = -q$, so that some care is needed in the choice of signs). Then

$$\alpha_1 = \tfrac{1}{2}(\beta + \gamma + \delta),$$

$$\alpha_2 = \tfrac{1}{2}(\beta - \gamma - \delta),$$

$$\alpha_3 = \tfrac{1}{2}(-\beta + \gamma - \delta),$$

and

$$\alpha_4 = \tfrac{1}{2}(-\beta - \gamma + \delta).$$

Notice that this means that $L = K(\beta, \gamma, \delta)$.

What are the possible Galois groups of an irreducible quartic? The exercises which follow provide an answer to this question.

Exercises

14.5 Suppose that G is a transitive subgroup of Σ_4. Show that G is either (i) Σ_4, (ii) A_4, (iii) the Vierergruppe N, (iv) cyclic of order 4 or (v) a non-abelian group of order 8, isomorphic to the group of rotations and reflections of a square.

14.6 Suppose that f is an irreducible quartic in $K[x]$ (where char K is not 2 or 3) and that $L : K$ is a splitting field extension for f. Let g be the cubic resolvent for f, and let M be a splitting field for g in L. Verify that the following table includes all possibilities and that it determines the Galois group of f in each case.

Discriminant	g	f	$\Gamma_K(f)$
No square root in K	Irreducible over K		Σ_4
Has square root in K	Irreducible over K		A_4
Has square root in K	Factorizes in $K[x]$		Vierergruppe
No square root in K	Factorizes in $K[x]$	Factorizes in $M[x]$	Cyclic of order 4
No square root in K	Factorizes in $K[x]$	Irreducible over M	Of order 8

14.7 Determine the Galois groups of the following quartics in $\mathbb{Q}[x]$:

   (i) $x^4 + 4x + 2$;
   (ii) $x^4 + 8x - 12$;
   (iii) $x^4 + 1$;
   (iv) $x^4 + x^3 + x^2 + x + 1$;
   (v) $x^4 - 2$.

15

Finite Fields

15.1 Finite Fields

In this chapter, we consider a field F with a finite number q of elements, and its Galois theory. Its prime subfield is isomorphic to $\mathbb{Z}_p$ for some prime number p, and we identify it with $\mathbb{Z}_p$. F is a finite-dimensional vector space over $\mathbb{Z}_p$, of dimension n, say, so that $q = p^n$.

Theorem 15.1 $F : \mathbb{Z}_p$ is the splitting field extension for the polynomial $x^q - x$, so that $F : \mathbb{Z}_p$ is normal, and any two fields of order q are isomorphic. The Galois group $\Gamma(L : \mathbb{Z}_p)$ is cyclic of order $q - 1$. If $G : \mathbb{Z}_p$ is a normal extension with $G \subset F$ then $[G : \mathbb{Z}_p]$ divides n, and if d divides n then there is a unique extension $G : \mathbb{Z}_p$ such that $G : \mathbb{Z}_p$ is normal.

Proof The multiplicative group $F^* = F \setminus \{0\}$ has $q - 1$ elements, so that $a^{q-1} = 1$ for each $a \in F^*$. Thus $a^q - a = 0$ for each of the q elements of F. Thus $x^q - x = \prod_{a \in F}(x - a)$, so that $F : \mathbb{Z}_p$ is the splitting field extension for $x^q - x$. Since splitting fields are essentially unique, any two fields with q elements are isomorphic. Let λ be the exponent of the group $\Gamma(F : \mathbb{Z}_p)$. Then every $a \in F^*$ is a root of $x^\lambda - 1$, so that $\lambda = q - 1$, and $\Gamma(F : \mathbb{Z}_p)$ is cyclic. The normal extensions $G : \mathbb{Z}_p$ with $G \subseteq F$ are therefore in bijective correspondence with the quotients of $\Gamma(F : \mathbb{Z}_p)$, which are in bijective correspondence with the divisors of n. $\qquad\square$

Can we find an interesting generator for the cyclic group $\Gamma(F : \mathbb{Z}_p)$?

Theorem 15.2 *Let* $\Phi(a) = a^p$. *Then* Φ *is an automorphism of* F *which generates* $\Gamma(F : \mathbb{Z}_p)$.

156

Proof If $a, b \in F$ then

$$(a+b)^p = a^p + \sum_{j=1}^{p-1} \binom{p}{j} a^j b^{p-j} + b^p = a^p + b^p,$$

since p divides the intermediate binomial coefficients. Since $(ab)^p = a^p b^p$, Φ is a monomorphism, and is therefore an automorphism of F. The set of elements fixed by Φ is a subfield G of F. If $a \in \mathbb{Z}_p$ then $\Phi(a) = a$, so that $\mathbb{Z}_p \subseteq G$. But the polynomial $x^p - x$ has only p roots in F, so that $G = \mathbb{Z}_p$. Thus it follows from the previous theorem that Φ generates $\Gamma(F : \mathbb{Z}_p)$. $\square$

Φ is called the *Frobenius automorphism*.

Exercises

15.1 Suppose that $|F| = q$, where $q = p^n$. How many elements of F^* generate the multiplicative group F^*?

15.2 Suppose that $|F| - q$, where $q = p^n$. What are the subfields of F? If G is a subfield of F, what is the Galois group $\Gamma(F : G)$?

15.2 Polynomials in $\mathbb{Z}_p[x]$

We have seen that any finite field is a splitting field for a polynomial $x^{q-1} - 1$ over a suitable prime subfield $\mathbb{Z}_p$. Let us now go in the opposite direction, considering polynomials in $\mathbb{Z}_p[x]$ and their splitting fields.

Let us start with the cyclotomic polynomial Φ_m over $\mathbb{Z}_p$. Let $F_m : \mathbb{Z}_p$ be a splitting field extension for Φ_m. Then $F_m : \mathbb{Z}_p$ is a splitting field extension for $x^m - 1$. Let E_m be the set of roots of $x^m - 1$ in F_m, and let P_m be the set of primitive roots of unity. Recall that there is a monomorphism Ψ of P_m onto U_m, the multiplicative group of units in $\mathbb{Z}_m$.

Proposition 15.3 *Suppose that p does not divide m. Then $x^m - 1$ is separable, and E_m is a cyclic subgroup of $(F_m^*, \times)$ of order m.*

Proof Since $D(x^m - 1) = mx^{m-1} \neq 0$, $x^m - 1$ is separable. If $a, b \in E_m$, then $(ab^{-1})^m = a^m b^{-m} = 1$, so that E_m is a subgroup of the cyclic group $(F_m^*, \times)$, and is therefore cyclic $\square$

What if p divides m?

Proposition 15.4 *Suppose that $m = lp^r$, where p does not divide l, $l > 1$ and $r > 0$. Then $F_m = F_l$, $E_m = E_l$ and each element of E_m has multiplicity p^r.*

Proof   This follows immediately from the fact that $x^m - 1 = (x^l - 1)^{p^r}$.   □

Suppose that p does not divide m. Let $\bar{p} = p \pmod{m}$. Then $\bar{p}$ is a unit in U_m; let $d(\bar{p}, m)$ be its order.

Theorem 15.5 *Suppose that p does not divide m. Then the cyclotomic polynomial Φ_m is the product of k irreducible polynomials in $\mathbb{Z}_p[x]$, each of degree $d(p, m)$, where $k = \phi(m)/d(p, m)$. $F_m : \mathbb{Z}_p$ is the splitting field for each of these polynomials.*

Proof   There exists $\eta \in P_m$ such that $\Psi(\eta) = \bar{p}$. Then η has degree $d(p, m)$, and its minimal polynomial g_η has degree $d(p, m)$. Since the Frobenius automorphism generates $\Gamma(F_m, \mathbb{Z}_p)$, it follows that $F_m = \mathbb{Z}_p(\eta)$, and $[F_m : \mathbb{Z}_p] = d(p, m)$. Thus $F_m : \mathbb{Z}_p$ is the splitting field for the polynomial g_η. Let R_η be the set of roots of g_η. If $\eta' \in P_m \setminus R(\eta)$, there exists $\sigma \in \Gamma(\Phi_m)$ such that $\sigma(\eta) = \eta'$, and then $R_{\eta'} = \sigma(R_\eta)$ and $g_{\eta'} = \prod_{\alpha \in R_\eta}(x - \sigma(\alpha))$. Thus Φ_m is the product of k irreducible polynomials of degree $d(p, m)$.   □

Corollary 15.6 *If m is a prime number, then Φ_m is irreducible in $\mathbb{Z}_p[x]$.*

Proof   For then $U_m = F_m *$.   □

Exercises

15.3 Suppose that $|F| = q$, where $q = p^n$. Show that if $a \neq 0$, then $x^p - x - a$ is irreducible over F. Show that it is irreducible over F_p if and only if it does not have a linear factor.

15.4 Suppose that $|F| - q$, where $q = p^n$ and that $f : F \to F$. Let $p(x) = \sum_{a \in F} f(a)(1 - (x - a)^{q-1})$ so that p is a polynomial of degree less than q. Show that $p(a) = f(a)$, for $a \in F$.

15.5 Suppose that $|F| = q$, where $q = p^n$. Show that $s_r = \sum_{a \in F} a^r = -1$ if $q - 1$ divides r, and that $s_r = 0$ otherwise.

15.6 Suppose that F is a finite field of order $q = p^n$, where p is an odd prime number, and that $\mathbb{Z}_p$ is its prime subfield. Show that $\mathbb{Z}_p = \{\alpha : \alpha^p = \alpha\}$. Show that $\alpha + \alpha^{-1} \in \mathbb{Z}_p$ if and only if $\alpha^{p+1} = 1$ or $\alpha^{P-1} = 1$ or $\alpha^{p-1} = 1$. Show that $\alpha^4 = -1$ if and only if $(\alpha + \alpha^{-1})^2 = 2$. Show that -1 is a square in $\mathbb{Z}_p$ if and only if $p = \pm 1 \bmod 8$.

15.3 Polynomials of Low Degree over a Finite Field

Let us see how the results that we have established apply to quadratic and cubic polynomials over a finite field K. Throughout this section, we shall suppose that K is a finite field of order p^n, where p is a prime number and n is a positive integer. Occasionally we shall require that p is not 2, or not 3. We shall see that the results depend on the nature of p^n. First let us consider quadratic polynomials. To begin with, we consider when -1 has a square root.

Proposition 15.7 *The polynomial $x^2 + 1$ has a root i in K if and only if $p^n = 4k + 1$ for some $k \in \mathbb{N}$.*

Proof -1 is a square if and only if K^* contains an element of order 4. The multiplicative group K^* is a cyclic group of order $p^n - 1$, and so it has an element of order 4 if and only if $4|(p^n - 1)$.   □

Thus $x^2 + 1$ factorizes in $K[x]$ if and only if $4|(p - 1)$ or n is even. If not, then the splitting field $K(i)$ for $x^2 + 1$ has order p^{2n}.

In order to consider a general quadratic polynomial over K, we complete the square, and consider a polynomial of the form $x^2 - \alpha$. Here we require that $p \neq 2$. If $x^2 + 1$ has a root in $K*$, then since the homomorphism $\beta \to \beta^2$ is 2 to 1, $x^2 - \alpha$ factorizes in $K[x]$ for half the values of α; for the other half, we must adjoin a square root of α. In contrast, suppose that $p^n = 4k - 1$ for some k, and that $x^2 - \alpha$ is irreducible over K. If we adjoin a square root i of -1, then $x^2 - \alpha$ splits over $K(i)$, since there is only one field of degree 2 over K; $K(i)$ is a splitting field for $x^2 - \alpha$.

Now let us consider cubic equations.

Proposition 15.8 *If $p^n = 3k + 1$, the polynomial $x^3 - 1$ splits in K and the equation $x^3 - \alpha = 0$ has a solution in K^* for one-third of the elements of K^*.*

If $p^n = 3k - 1$ for some $k \in \mathbb{N}$, the polynomial equation $x^n - \alpha = 0$ has a unique solution β in K^* for every $\alpha \in K^*$, $x^3 - \alpha = (x - \beta)(x^2 + \beta x + \beta^2)$, and $x^2 + \beta x + \beta^2$ is irreducible in $K[x]$.

Proof Let $\phi(\beta) = \beta^3$ for $\beta \in K^*$. ϕ is a homomorphism of K^* into itself. K^* has three elements of order 3 if and only if the kernel J of the homomorphism $\phi(x) = x^3$ has order divisible by 3, which happens if and only if $3|(p^n - 1)$. In this case, $|\phi(K^*)| = |K^*|/|J| = |K^*|/3$.

If $p^n = 3k - 1$, the kernel of ϕ is $\{1\}$, so that ϕ is an isomorphism of K^* onto K^*; thus there is just one β such that $\beta^3 = \alpha$. Since $x^3 - \alpha = (x - \beta)(x^2 + \beta x + \beta^2)$, $x^2 + \beta x + \beta^2$ must be irreducible in $K[x]$.   □

With this information, we can now use Cardan's method to solve a general cubic equation; here we require that $p \neq 2$ or 3.

Similarly, we can now follow the arguments for an irreducible quartic polynomial f in $K[x]$. In fact, the argument is a little simpler, since the Galois group Γ_f is cyclic.

Of course, the Galois group of any polynomial in $K[x]$ is cyclic, and the polynomial is therefore always soluble by radicals, but we shall go no further in this direction.

Exercises

15.7 Suppose that $|K| = 2^n$. How many irreducible monic quadratic polynomials in $K[x]$ are there?

15.8 Suppose that $|K| = 4$. Construct addition and multiplication tables for K. Which monic quadratic polynomials in $K[x]$ are not irreducible?

15.9 Suppose that $|K| = 4$, with elements $0, 1, \alpha, \beta$, and that f and g are two irreducible monic quadratic polynomials in $K[x]$. Construct splitting fields for f and g, and find an isomorphism between them.

16

Quintic Polynomials

What about quintic polynomials? There are quintic polynomials in $\mathbb{Z}[x]$ which cannot be solved by adjoining radicals. This was proved by Abel in 1824, some five years before the fundamental work of Galois. But Galois theory helps explain why this is so: we have seen, using Theorem 9.7, that if $f(x) = x^5 - 4x + 2 \in \mathbb{Z}(x)$ then $\Gamma(f) = \Sigma_5$. Σ_5 is not soluble, and so is not soluble by radicals.

Let us now consider the possible Galois groups of irreducible quintics. Suppose that f is an irreducible separable quintic in $K[x]$. If f is solvable by radicals then $\Gamma_K(f)$ is isomorphic either to W, which has order 20, or to D_{10}, the group of rotations and reflections of a regular pentagon, which has order 10, or to the cyclic group $\mathbb{Z}_5$ of order 5. If f is not solvable by radicals then $\Gamma_K(f)$ is isomorphic to the alternating group A_5 or the full symmetric group Σ_5.

Let us list some examples of irreducible quintic polynomials in $\mathbb{Z}[x]$, together with their Galois groups, to show that all possibilities can occur:

(a) $x^5 + x^4 - 4x^3 - 3x^2 + 3x + 1$ $\qquad\qquad\qquad\qquad$ $\mathbb{Z}_5$

(b) $x^5 - 5x + 12$ $\qquad\qquad\qquad\qquad\qquad\qquad\qquad$ D_{10}

(c) $x^5 - 2$ $\qquad\qquad\qquad\qquad\qquad\qquad\qquad\qquad\quad$ W

(d) $x^5 + 20x + 16$ $\qquad\qquad\qquad\qquad\qquad\qquad\qquad$ A_5

(e) $x^5 - 4x + 2$ $\qquad\qquad\qquad\qquad\qquad\qquad\qquad\quad$ Σ_5

We have discussed case (e), and case (c) follows from the discussion of splitting fields: if $[L : \mathbb{Q}]$ is a splitting field for $x^5 - 2$ then $L = \mathbb{Q}(\alpha, \beta)$, where $\alpha^5 = 2$ and $\beta - e^{2\pi i/5}$ is a root of the quartic cyclotomic polynomial $\Sigma_{n=0}^{4} x^n$, so that $|L : \mathbb{Q}| = 20$, and the Galois group has 20 elements.

For case (a), let us start with the cyclotomic polynomial $\sum_{n=0}^{1} x^n$, which has the 10 roots $\{\alpha^n : 1 \leq n \leq 10\}$ where $\alpha = e^{2\pi/11}$ is an 11th root of unity. By Theorem 9.7, this has a cyclic Galois group of order 10, generated by g, say. Then g^5 has order 2, and is the mapping $g^5(\alpha^r) = \alpha^{-r} = \bar{\alpha}^r$. Then if $\beta = \alpha + \bar{\alpha} = 2\cos(2\pi/11)$, $\beta \in \mathbb{R}$ and $|\mathbb{Q}(\beta) : \mathbb{Q}| = 5$. Example (a) is the minimal polynomial of β, which therefore has $\mathbb{Z}_5$ as its Galois group.

In general, the calculation of Galois groups is difficult. In fact, if $f \in \mathbb{Z}[x]$ is irreducible and $\bar{f}$ is the corresponding polynomial in $\mathbb{Z}_p$ (where p is a prime number), then the Galois group of $\bar{f}$ is isomorphic to a subgroup of the Galois group of f. This enables us to deal with case (d): the discriminant of the polynomial is $2^{16}5^6$, so that the Galois group is contained in A_5. The corresponding polynomial in $\mathbb{Z}_7$ is

$$x^5 - x + 2 = (x+2)(x+3)(x^3 + 2x^2 - 2x - 2)$$

where the cubic factor is irreducible, and so its Galois group contains a cycle of order 3. Thus the Galois group of the original polynomial must be A_5.

What about the polynomial f of case (b). Its discriminant is $2^{12} \cdot 5^6$, and the corresponding polynomial in $\mathbb{Z}_3$ is $x(x^2 + x - 1)(x^2 - x - 1)$. This implies that the Galois group of f is either D_{10} or A_5. How to discriminate between these? One way to proceed is the following. Let $\alpha_1, \ldots, \alpha_5$ be the roots of the polynomial in a splitting field L. We consider the 10 elements $\alpha_i + \alpha_j$ of L, with $1 \leq i < j \leq 5$. It is easy to verify that these are distinct, so that the polynomial

$$g = \prod_{1 \leq i < j \leq 5} (x - (\alpha_i + \alpha_j))$$

has 10 distinct roots. g is invariant under $\Gamma_{\mathbb{Q}}(f)$, and so $g \in \mathbb{Q}[x]$. It is clear that g splits over L. Since

$$\alpha_1 = \left(\sum_{i=1}^{5} \alpha_i\right) - (\alpha_2 + \alpha_3) - (\alpha_4 + \alpha_5)$$

and since similar equations are satisfied by α_2, α_3, α_4 and α_5 it follows that $L : \mathbb{Q}$ is a splitting field extension for g. Suppose that f had Galois group A_5. Then the Galois group would act transitively on the roots of g, and so g would be irreducible. Using a computer it can be shown that this is not so.

The calculation of Galois groups is not easy! It is even more difficult to determine whether or not a given finite group can be a Galois group. I am grateful to Leonard Soicher for showing me examples (b) and (d).

Exercises

16.1 Suppose that f is a quintic in $\mathbb{Q}[x]$ whose Galois group contains D_{10}. Show that the 10 elements $\alpha_i + \alpha_j$ $(1 \leqslant i < j \leqslant 5)$ are distinct (where $\alpha_1, \ldots, \alpha_5$ are the roots of f in $\mathbb{C}$).

16.2 Express $\cos(5\theta)$ as a polynomial in $\cos(\theta)$. Is $\cos(\pi/10)$ solvable by radicals?

17

Further Theory

We now establish some further results, of general theoretical interest.

Suppose that $f \in K[x]$ has Galois group $\Gamma_K(f)$, and that $L : K$ is an extension. Then we can consider f as an element of $L[x]$, and can consider the Galois group $\Gamma_L(f)$. In each case, we can consider the Galois group as a permutation of the roots of f. In the first case we must fix K, and in the second we must fix the larger field L. Thus we should expect $\Gamma_L(f)$ to be a subgroup of $\Gamma_K(f)$.

For example, suppose that f is separable over K, that $F : K$ is a splitting field extension for f over K and that $F : L : K$. Then

$$\Gamma_L(f) = \Gamma(F : L) \subseteq \Gamma(F : K) = \Gamma_K(f).$$

In general, L is not an intermediate field: the *theorem on natural irrationalities* says that in fact this does not affect things.

Theorem 17.1 *Suppose that $f \in K[x]$ and that $L : K$ is an extension. Let $N : L$ be a splitting field extension for f over L, let $\alpha_1, \ldots, \alpha_n$ be the roots of f in N and let $M = K(\alpha_1, \ldots, \alpha_n)$ (so that $M : K$ is a splitting field extension for f over K). Let L_0 be the fixed field of $\Gamma_L(f)$. Then if $\sigma \in \Gamma_L(f)$, $\sigma|_M \in \Gamma(M : L_0 \cap M)$, and the map $\theta : \sigma \to \sigma|_M$ is an isomorphism of $\Gamma_L(f)$ onto the subgroup $\Gamma(M : L_0 \cap M)$ of $\Gamma_K(f)$.*

Proof   We have the following diagram of inclusions:

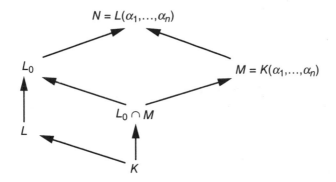

If $\sigma \in \Gamma(N : L)$, σ fixes K and permutes $\{\alpha_1, \ldots, \alpha_n\}$ and so $\sigma(M) \subseteq M$. Thus $\sigma|_M$ is an automorphism of M, which clearly fixes $L_0 \cap M$. Since the group multiplication is the composition of mappings, θ is a homomorphism of $\Gamma(N : L)$ into $\Gamma(M : L_0 \cap M)$.

If $\theta(\sigma)$ is the identity, σ fixes $\alpha_1, \ldots, \alpha_n$ and so (since σ fixes L) σ must be the identity on $N = L(\alpha_1, \ldots, \alpha_n)$. Thus θ is one-to-one.

Let V be the fixed field of $\theta(\Gamma(N : L))$. As we have seen, $V \supseteq L_0 \cap M$. Suppose that $x \in M$ and $x \notin L_0 \cap M$. Since L_0 is the fixed field of $\Gamma(N : L)$, there exists $\sigma \in \Gamma(N : L)$ such that $\sigma(x) \neq x$. Thus $\theta(\sigma)(x) \neq x$, and so $x \notin V$. Thus $V = L_0 \cap M$, and so, by Theorem 9.3,

$$\theta(\Gamma(N : L)) = \Gamma(M : V) = \Gamma(M : L_0 \cap M).$$

Note that if f is separable over K, then f is separable over L, and so $L = L_0$: in this case the theorem becomes a little simpler.   □

17.1 Simple Extensions

Theorem 17.2 *An algebraic extension $L : K$ is simple if and only if there are only finitely many intermediate fields.*

Proof   First suppose that there are only finitely many intermediate fields. $L : K$ must be finitely generated over K, for otherwise there is a strictly increasing infinite sequence of intermediate fields. Thus $L : K$ is finite (Theorem 4.7). If K is finite, L is finite, and so $L : K$ is simple (Exercise 10.1). We may therefore restrict attention to the case where K is *infinite*.

Let $F_1, \ldots, F_k$ be the set of proper intermediate fields. They are proper linear subspaces of L, and so $\cup F_{j=1}^k F_j \neq L$ by Theorem 3.1. Let $\alpha \in L \setminus \cup F_{j=1}^k F_j$. Then $K(\alpha)$ is a subfield of L which must equal L.

Conversely, suppose that $L = K(\alpha)$ is simple and algebraic over K. Let m be the minimal polynomial for α over K. m is irreducible over K, but of course m factorizes over L. Nevertheless, m has only finitely many monic divisors $d_1, \ldots, d_k$, say, in $L[x]$. Now suppose that F is an intermediate field. Let m_F be the minimal polynomial for α over F. Considering m as an element of $F[x]$, we see that $m_F | m$. But this means that $m_F | m$ in $L[x]$, and so $m_F = d_i$ for some $1 \leqslant i \leqslant k$. The proof will therefore be complete if we can show that m_F determines F. Let

$$m_F = a_0 + a_1 x + \cdots + a_r x^r,$$

and let $F_0 = K(a_0, \ldots, a_r)$. Then $F_0 \subseteq F$, and so m_F is irreducible over F_0. Thus m_F is the minimal polynomial for α over F_0. As $L = F_0(\alpha)$, $[L : F_0] = $ degree m_F (Theorem 4.5). But $[L : F] = $ degree m_F, by the same argument, so that as $F \supseteq F_0$ we must have $F = F_0 = K(a_0, \ldots, a_r)$. Thus m_F determines F; this completes the proof. $\square$

Exercises

17.1 Use Exercise 3.1 to give another proof that $L : K$ is simple if K is infinite and there are only finitely many intermediate fields.

17.2 Suppose that $K(t) : K$ is a simple transcendental extension. Show that there are infinitely many intermediate fields.

17.2 The Theorem of the Primitive Element

Theorem 17.3 (The theorem of the primitive element) *Suppose that $L : K$ is finite and separable. Then there are only finitely many intermediate fields, so that $L : K$ is simple.*

Proof Suppose that $\alpha_1, \ldots, \alpha_n$ generate L over K. Let $g = m_{\alpha_1} \ldots m_{\alpha_n}$, where m_{a_i} is the minimal polynomial of α_i over K. Then g is separable over K. Let $N : L$ be a splitting field extension for g over L. As $\alpha_1, \ldots, \alpha_n$ are roots of g, $N : K$ is also a splitting field extension for g over K. Thus $N : K$ is normal (Corollary 1 to Theorem 7.1) and separable (Corollary 10.3 to Theorem 8.3) and is therefore a Galois extension. Thus K is the fixed field of $\Gamma(N : K)$.

Now $\Gamma(N : K)$ is finite, and so it has finitely many subgroups. By the fundamental theorem of Galois theory, these are in one-to-one correspondence with the fields intermediate between N and K. Thus there are finitely many fields intermediate between N and K, and *a fortiori* there are finitely many fields intermediate between L and K.   □

Corollary 17.4 *Suppose that $L : K$ is a Galois extension and that $[L : K] = k$. Then $[L : K]$ is simple; if $L = K(\alpha)$ and m is the minimum polynomial for α over K, then $L : K$ is a splitting field extension for m over K, and $1, \alpha, \ldots, \alpha^{k-1}$ is a basis for L over K.*

Proof For then $g = m$ and $L = N$. If j is the degree of m then $[K(\alpha) : K] = j$, and so $j = k$. Similarly, $1, \alpha, \ldots, \alpha^{j-1}$ are linearly independent over K, and so form a basis for $K(\alpha) = L$ over K.   □

Let us consider a very easy example. $\mathbb{Q}(\sqrt{2}, \sqrt{3}) : \mathbb{Q}$ is a splitting field extension for $f = x^4 - 5x^2 + 6 = (x^2 - 2)(x^2 - 3)$ over $\mathbb{Q}$. $[\mathbb{Q}(\sqrt{2}, \sqrt{3}) : \mathbb{Q}] = 4$, and the Galois group $\Gamma_{\mathbb{Q}}(f)$ is best described by its action on $\sqrt{2}$ and $\sqrt{3}$:

	$\sigma_0 = e$	σ_1	σ_2	σ_3
$\sqrt{2}$	$\sqrt{2}$	$-\sqrt{2}$	$\sqrt{2}$	$-\sqrt{2}$
$\sqrt{3}$	$\sqrt{3}$	$\sqrt{3}$	$-\sqrt{3}$	$-\sqrt{3}$

$\Gamma_{\mathbb{Q}}(f)$ is isomorphic to $\mathbb{Z}_2 \times \mathbb{Z}_2$ and has three non-trivial subgroups: $\{\sigma_0, \sigma_1\}$, $\{\sigma_0, \sigma_2\}$ and $\{\sigma_0, \sigma_3\}$. The corresponding fixed fields are $\mathbb{Q}(\sqrt{3})$, $\mathbb{Q}(\sqrt{2})$ and $\mathbb{Q}(\sqrt{6})$. If α is any element of $\mathbb{Q}(\sqrt{2}, \sqrt{3})$ which does not belong to any of these three intermediate fields, then $\mathbb{Q}(\sqrt{2}, \sqrt{3}) = \mathbb{Q}(\alpha)$.

Exercises

17.3 Let p be a prime, let $J = \mathbb{Z}_p(\alpha)$, where α is transcendental over $\mathbb{Z}_p$, and let $K = J(\beta)$, where β is transcendental over J. Let $L : K$ be a splitting field extension for $(x^p - \alpha)(x^p - \beta)$.

    (i) Show that $[L : K] = p^2$.

    (ii) Show that if $\gamma \in L$ then $\gamma^p \in K$.

    (iii) Show that $L : K$ is not simple.

    (iv) In the case where $p = 2$, find all the intermediate fields $L : M : K$.

17.4 Suppose that $L : K$ is a finite separable extension and that $M : L$ is a finite simple extension. Show that $M : K$ is a simple extension.

17.3 The Normal Basis Theorem

We need a preliminary result.

Theorem 17.5 *Suppose that R is an infinite subset of an integral domain S and that f is a non-zero element of $S[x_1, \ldots, x_n]$. Then there exists $(r_1, \ldots, r_n)$ in R^n such that $f(r_1, \ldots, r_n) \neq 0$.*

Proof We prove this by induction on n. In the case where $n = 1$, f has only finitely many roots in F, the field of fractions of S, and so f has only finitely many roots in S: since R is infinite, there exists r in R such that $f(r) \neq 0$.

Suppose that the result is true for $n - 1$. $S[x_1]$ is an infinite integral domain, and we can consider f as a non-zero element of $S[x_1][x_2, \ldots, x_n]$. By the inductive hypothesis, there exist $r_2, \ldots, r_n$ in R such that $f(x_1, r_2, \ldots, r_n)$ is a non-zero element of $S[x_1]$; by the case where $n = 1$, there exists r_1 in R such that $f(r_1, \ldots, r_n) \neq 0$.   □

This theorem requires two separate proofs, in the cases where the underlying field is finite or infinite.

Theorem 17.6 (The normal basis theorem) *First suppose that K is infinite. Suppose that $L : K$ is a Galois extension, with Galois group $G = \{\sigma_1, \ldots, \sigma_n\}$. Then there exists l in L such that $(\sigma_1(l), \ldots, \sigma_n(l))$ is a basis for L over K.*

Proof By relabelling G if necessary, we can suppose that σ_1 is the identity. Let us define $p(i, j)$ by the formula

$$\sigma_i \sigma_j = \sigma_{p(i,j)}$$

for $1 \leqslant i \leqslant n$, $1 \leqslant j \leqslant n$. Let $x_1, \ldots, x_n$ be indeterminates and let M be the $n \times n$ matrix $(x_{p(i,j)})_{i=1, j=1}^{n, \ n}$ with entries in $K[x_1, \ldots, x_n]$. Let $f = \det M$. Then $f \in K[x_1, \ldots, x_n]$, and f is non-zero, since x_1 occurs once in each row and once in each column, and so the coefficient of x_1^n is 1 or -1.

Now let $(b_1, \ldots, b_n)$ be a basis for L over K. By Theorem 9.2 the n trajectories

$$(T(b_j))_{j=1}^n = ((\sigma_i(b_j))_{i=1}^n)_{j=1}^n$$

are linearly independent over L in L^n: in other words the $n \times n$ matrix $(\sigma_i(b_j))_{i=1, j=1}^{n, \ n}$ is invertible; let $C = (c_{ij})$ be its inverse.

We now set

$$g(x_1, \ldots, x_n) = f\left(\sum_j \sigma_1(b_j)x_j, \ldots, \sum_j \sigma_n(b_j)x_j\right).$$

Since

$$f(x_1, \ldots, x_n) = g\left(\sum_j c_{1j}x_j, \ldots, \sum_j c_{nj}x_j\right),$$

g is a non-zero element of $L[x_1, \ldots, x_n]$. This means that there exist $k_1, \ldots, k_n$ in K such that $g(k_1, \ldots, k_n) \neq 0$. We set $l = k_1 b_1 + \cdots + k_n b_n$. Then

$$0 \neq g(k_1, \ldots, k_n) = f \left(\sum_j \sigma_1(b_j)k_j, \ldots, \sum_j \sigma_n(b_j)k_j \right)$$

$$= f(\sigma_1(l), \ldots, \sigma_n(l))$$

$$= \det((\sigma_{p(i,j)}(l))) = \det((\sigma_i(\sigma_j(l)))).$$

This means that the matrix $(\sigma_i(\sigma_j(l)))$ is invertible, and so by Theorem 9.2 $(\sigma_1(l), \ldots, \sigma_n(l))$ is a basis for L over K. $\square$

Let us now prove the normal basis theorem for finite fields.

Suppose that K is a finite field and that $L : K$ is a finite extension of degree n. Then $\Gamma(L : K)$ is a cyclic group of order n, generated by $T = \Phi_n$, the Frobenius automorphism, whose minimal polynomial is $m_T(x) = x^n - 1$.

If $\lambda \in L$ let $I_\lambda = \{f \in K[x] : f(T)\lambda = 0$. I_λ is an ideal in $K[x]$, and so $I_\lambda = < m_\lambda >$, for some $m_\lambda \in K[x]$, since $K[x]$ is a principal ideal domain. m_λ is the minimal polynomial for λ, and clearly $m_\lambda | m_T$.

Lemma 17.7 *(i) m_T is the lowest common multiple of $\{m_\lambda : \lambda \in L\}$.*
(ii) If $\mu = f(T)\lambda$ then $m_\mu | m_\lambda$.
(iii) If m_λ and m_μ are coprime, then $m_{\lambda+\mu} = m_\lambda m_\mu$.

Proof   (i) Let m be the lowest common multiple of $\{m_\lambda : \lambda \in L\}$. Then clearly $m_T | m$. But also $m(T)\lambda = 0$ for all λ, so that $m | m_T$.
(ii) $m_\lambda(T)\mu = m_\lambda(T)f(T)\lambda = f(T)m_\lambda(T)\lambda = 0$, so that $m_\mu | m_\lambda$.
(iii) Clearly $m_{\lambda+\mu} | m_\lambda m_\mu$.

Let $v = m_\lambda(T)(\lambda + \mu) = m_\lambda(T)\mu$. Then $m_v | m_{\lambda+\mu}$, by (ii). Then $0 = m_v(T)v = m_v(T)m_\lambda(T)\mu$, so that $m_\mu | m_v m_\lambda$. Since m_λ and $m_m u$ are coprime, $m_\mu | m_v$ and so $m_\mu | m_{\lambda+\mu}$. Similarly $m_\lambda | m_{\lambda+\mu}$. Thus $m_\lambda m_\mu | m_{\lambda+\mu}$, and so $m_{\lambda+\mu} = m_\lambda m_\mu$. $\square$

We now complete the proof. Suppose that $m = p_1^{r_1} \cdots p_k^{r_k}$, where $p_1, \ldots, p_k$ are distinct prime numbers. By (ii), for $1 \leq i \leq k$ there exists λ_i such that $m_{\lambda_i} = p_i^{r_i} f_i$, so that $\mu_i = f_i(T)\lambda_i$ has minimal polynomial $p_i^{r_i}$. Let $\mu = m_1 + \cdots + m_k$. Then $m_\mu(x) = m_T(x) = x^n - 1$. Thus $\{\mu.T(\mu), \ldots, T^{n-1}(\mu)\}$ is a normal basis.

Exercise

17.5 Show that the primitive nth roots of unity over $\mathbb{Q}$ form a normal basis for the splitting field of $x^n - 1$ over $\mathbb{Q}$ if and only if n has no repeated prime factors.

18

The Algebraic Closure of a Field

18.1 Introduction

If $f \in \mathbb{Q}[x]$ we can consider f as an element of $\mathbb{C}[x]$, and then f splits over $\mathbb{C}$. We therefore have the comforting conclusion that, whenever $f \in \mathbb{Q}[x]$, we can find a splitting field extension for f which is a subfield of the fixed field $\mathbb{C}$.

In this chapter we shall show that a similar phenomenon occurs for any field K. We must make some definitions. A field L is said to be *algebraically closed* if every f in $L[x]$ splits over L. Thus the 'fundamental theorem of algebra' states that $\mathbb{C}$ is algebraically closed. An extension $L : K$ is called an *algebraic closure* of K if $L : K$ is algebraic and L is algebraically closed. Note that $\mathbb{C} : \mathbb{Q}$ is *not* an algebraic closure of $\mathbb{Q}$ since $\mathbb{C} : \mathbb{Q}$ is not algebraic (Exercise 4.17).

The next theorem gives two useful characterizations of an algebraic closure:

Theorem 18.1 *Suppose that* $L : K$ *is an extension. The following are equivalent:*

(i) $L : K$ *is an algebraic closure of* K.
(ii) $L : K$ *is algebraic, and every irreducible* f *in* $K[x]$ *splits over* L.
(iii) $L : K$ *is algebraic, and if* $L' : L$ *is algebraic then* $L = L'$.

Proof Clearly (i) implies (ii). Suppose that (ii) holds and that $L' : L$ is algebraic. Then $L' : K$ is also algebraic (Theorem 4.8). Suppose that $\alpha' \in L'$. Let m be the minimal polynomial of α' over K. Then m is irreducible and so, by hypothesis, m splits over L:

$$m = (x - \lambda_1) \ldots (x - \lambda_n).$$

As $m(\alpha') = 0$, $\alpha' = \lambda_j$ for some j, and so $\alpha' \in L$. Thus $L = L'$ and (iii) holds. Finally suppose that (iii) holds, and that $f \in L[x]$. By Theorem 6.3, there is a splitting field extension L' for f over L. $L' : L$ is algebraic, by the corollary to Theorem 6.1 and so, by hypothesis, $L' = L$. Thus f splits over L, and so L is algebraically closed. Consequently $L : K$ is an algebraic closure of K. $\square$

Corollary *Suppose that $L : K$ is an extension and that L is algebraically closed. Let L_a be the field of elements of L which are algebraic over K. Then $L_a : K$ is an algebraic closure of K.*

In particular, if A is the field of complex numbers which are algebraic over $\mathbb{Q}$, then $A : \mathbb{Q}$ is an algebraic closure for $\mathbb{Q}$.

18.2 The Existence of an Algebraic Closure

We now turn to the fundamental theorem concerning algebraic closures.

Theorem 18.2 *If K is a field, there exists an algebraic closure $L : K$.*

The generality of this statement suggests that we may need to use the axiom of choice, and the maximal nature of an algebraic closure revealed by Theorem 18.1 reinforces this belief. It is, however, necessary to proceed with some care. Let us begin by giving a *fallacious* argument.

Partially order the algebraic extensions $M : K$ by saying that $M_1 : K \geq M_2 : K$ if $M_2 : K$ is a subfield of M_1. If $\mathscr{C}$ is a chain of extensions $M : K$, let $N = \cup\{M : M : K \in \mathscr{C}\}$. If $\alpha, \beta \in N$, there exists $M : K$ in $\mathscr{C}$ such that α and β are in M. Define $\alpha\beta, \alpha + \beta$ and α^{-1} (if $\alpha \neq 0$) by the operations in M. This does not depend on M, and so N is a field, and $N : K$. If $\alpha \in N$, $\alpha \in M$ for some M, and so α is algebraic over K. Thus $N : K$ is an upper bound for $\mathscr{C}$. By Zorn's lemma, there is a maximal algebraic extension, and by Theorem 18.1, this is an algebraic closure.

What is wrong with this argument? The error comes at the very beginning, when we try to compare extensions. Recall that an extension is really a triple (i, K, M), where i is a monomorphism from K into M. Thus in general we cannot compare extensions in the way that is suggested.

Nevertheless, the fallacious argument has some virtue, and it is possible, by considering fields which, as sets, are subsets of a sufficiently large fixed set, to produce a correct argument along the lines which the fallacious argument suggests. Exercises 18.1 and 18.2 show one way in which this can be done. We shall instead give a more 'ring-theoretic' argument, which uses the axiom of choice by appealing to Theorem 2.16.

We consider a ring of polynomials in very many variables. If f is a non-constant monic polynomial in $K[x]$ of degree n, then f has at most n roots in a splitting field extension: we introduce an indeterminate to correspond to each of these possible roots. Let U be the set of all pairs (f, j), where f is a non-constant monic polynomial in $K[x]$ and $1 \leq j \leq$ degree f. For each (f, j) in U, we introduce an indeterminate $x_j(f)$, and consider the polynomial ring $K[X_U]$ of polynomials with coefficients in K and with indeterminates

$$X_U = \{x_j(f): (f, j) \in U\}.$$

Now suppose that f is a non-constant monic polynomial in $K[x]$. We can write

$$f = x^n - a_1(f)x^{n-1} + \cdots + (-1)^n a_n(f)$$

(notice that we have not written monic polynomials in this form before: as we shall see, this can be a very useful form to use). Let $g(f)$ be the element of $K[X_U][x]$ that has $x_1(f), \ldots, x_n(f)$ as roots:

$$g(f) = \prod_{j=1}^{n}(x - x_j(f))$$

$$= x^n - s_1(f)x^{n-1} + \cdots + (-1)^n s_n(f),$$

where

$$s_j(f) = \sum_{i_1 < \cdots < i_j} x_{i_1}(f) \ldots x_{i_j}(f) \in K[X_U]$$

is the jth elementary symmetric polynomial in $x_1(f), \ldots, x_n(f)$.

The idea of the proof is to identify f and $g(f)$, and to exploit the fact that $g(f)$ splits in $K[X_U][x]$. With this in mind, we set

$$t_i(f) = s_i(f) - a_i(f)$$

for $1 \leq i \leq n$. Let I be the ideal in $K[X_U]$ generated by all the elements $t_i(f)$ as f and i vary. The main step in the proof is to show that I is a proper ideal in $K[X_U]$. For this, it is sufficient to show that $1 \notin I$; in other words to show that it is not possible to find $r_1, \ldots, r_N$ in $K[X_U]$ and elements $t_{i_1}(f_1), \ldots, t_{i_N}(f_N)$ such that

$$1 = r_1 t_{i_1}(f_1) + \cdots + r_N t_{i_N}(f_N).$$

Suppose that such an expression were to exist. By Theorem 6.3, there exists a splitting field $L : K$ for the polynomial $h = f_1 \ldots f_N$. Then each f_k splits over L; we can write

$$f_k = (x - \alpha_1(k)) \ldots (x - \alpha_{n_k}(k))$$

where $n_k = $ degree f_k. Note that

$$a_j(f_k) = \sum_{i_1 < \cdots < i_j} \alpha_{i_1}(k) \ldots \alpha_{j_J}(k).$$

We now consider the evaluation map E from $K[X_U]$ to L which sends $X_i(f_k)$ to $\alpha_i(k)$ for $1 \leq i \leq n_k$ and $1 \leq k \leq N$, and which sends all the other indeterminates to 0. Then $E(s_j(f_k)) = a_j(f_k)$, and so it follows from the definition of $t_i(f)$ that

$$E(t_i(f_k)) = 0 \quad \text{for } 1 \leq i \leq n_k, 1 \leq k \leq N,$$

so that

$$1 = E(1) = E(r_1)E(t_{i_1}(f_1)) + \cdots + E(r_N)E(t_{i_N}(f_N)) = 0.$$

This gives the contradiction that we are looking for.

Since I is a proper ideal of $K[X_U]$, there exists a maximal proper ideal J of $K[X_U]$ which contains I, by Theorem 2.16. (This is where we use the axiom of choice.) By Theorem 2.35, $K[X_U]/J$ is a field, M say. We now let $j = qi$, where i is the natural monomorphism from K into $K[X_U]$, and q is the quotient map from $K[X_U]$ onto M:

$$K \xrightarrow{i} K[X_U] \xrightarrow{q} M.$$

Then (j, K, M) is an extension of K. Let us set $\beta_j(f) = q(x_j(f))$, for all $u = (f, j) \in U$.

Now suppose that

$$f = x^n - a_1(f)x^{n-1} + \cdots + (-1)^n a_n(f)$$

is a non-constant monic polynomial in $K[x]$. Then

$$j(f) = x^n - j(a_1(f))x^{n-1} + \cdots + (-1)^n j(a_n(f))$$

is the corresponding polynomial in $M[x]$. But

$$j(a_k(f)) = q(i(a_k(f))) = q(s_k(f)),$$

since $s_k(f) - i(a_k(f)) = t_k(f) \in I \subseteq J$. Thus

$$j(f) = x^n - q(s_1(f))x^{n-1} + \cdots + (-1)^n q(s_n(f))$$
$$= q(x^n - s_1(f)x^{n-1} + \cdots + (-1)^n s_n(f))$$
$$= q((x - x_1(f))(x - x_2(f)) \ldots (x - x_n(f)))$$
$$= (x - \beta_1(f)) \ldots (x - \beta_n(f)),$$

and $j(f)$ splits over M. Further, each $\beta_k(f)$ is algebraic over $j(K)$ (since it is a root of $j(f)$) and the $\beta_k(f)$ generate M over K: thus $M : K$ is algebraic, by Corollary 2 to Theorem 4.7. Consequently (j, K, M) is an algebraic closure of K. If K is countable, then $K(X_U)$ is countable, so that by using Theorem A.1 we can avoid using the axiom of choice.

Exercises

18.1 (i) Suppose that U is a non-empty set, and that $P(U)$ is the set of subsets of U. Show that if $V \subseteq U$ and $f : V \to P(U)$ is a mapping, then f is not onto. (Consider $\{x : x \in V, x \notin f(x)\}$.)

   (ii) Suppose that U is a non-empty set and that $V \subseteq W \subseteq U$. Show that if $f : V \to P(U)$ is one-to-one then there exists a one-to-one map $g : W \to P(U)$ such that $g|_v = f$. (Use Zorn's lemma.)

18.2 Suppose that K is a field. Let $U = K[x] \times \mathbb{Z}^+$.

   (i) Show that if (k, K, L) is an algebraic extension, then there exists a one-to-one mapping of L into U. (Use Zorn's lemma.)

   (ii) Suppose that (k, K, L) and (l, L, M) are algebraic extensions. Show that if $f : L \to P(U)$ is one-to-one then there exists a one-to-one map $g : M \to P(U)$ such that $f = gl$.

18.3 Suppose that K is a field. Let $U = K[x] \times \mathbb{Z}^+$.

   (i) If $\alpha \in K$, let $j(\alpha) = \{(x - \alpha, 1)\}$. Show that $j : K \to P(U)$ is one-to-one and that $j(K)$ can be given the structure of a field in such a way that j is a field isomorphism.

   (ii) Let $\mathscr{F}$ be the set of triples $(S, +, .)$ where

      (a) $j(K) \subseteq S \subseteq P(U)$;
      (b) $(S, +, .)$ is a field, $F(S)$ say;
      (c) $(i, j(K), F(S))$ is an algebraic extension (here i is the inclusion mapping).

   Define a partial order on $\mathscr{F}$ by saying that $(S_1, +_1, \cdot_1) \leq (S_2, +_2, \cdot_2)$ if $S_1 \subseteq S_2$ and $(i, F(S_1), F(S_2))$ is an extension (again, i is the inclusion mapping). Show that under this order, $\mathscr{F}$ has a maximal element (Zorn's lemma).

   (iii) Use Theorem 6.2 or 6.3 to show that if $(S, +, .)$ is a maximal element of $\mathscr{F}$ then $(j, K, F(S))$ is an algebraic closure for K. (Here j is considered as a mapping of K into $F(S)$.)

18.3 The Uniqueness of an Algebraic Closure

We now consider problems of uniqueness. First we establish an extension theorem: this uses Zorn's lemma in a very standard way.

Theorem 18.3 *Suppose that* $i\colon K_1 \to K_2$ *is a monomorphism, that* $L\colon K_1$ *is algebraic and that* K_2 *is algebraically closed. Then there exists a monomorphism* $j\colon L \to K_2$ *such that* $j|_{K_1} = i$.

Proof   Let S denote all pairs (M, θ), where M is a subfield of L containing K_1, and θ is a monomorphism from M into K_2 such that $\theta|_{K_1} = i$. Partially order S by setting $(M_1, \theta_1) \le (M_2, \theta_2)$ if $M_1 \subseteq M_2$ and $\theta_2|_{M_1} = \theta_1$. If $\mathscr{C}$ is a chain in S, let $N = \cup\{M\colon (M, \theta) \in \mathscr{C}\}$. If $n \in N$, then $n \subset M$ for some $(M, \theta) \in \mathscr{C}$. Set $\phi(n) = \theta(n)$. It is now straightforward to verify that ϕ is well defined, that $\phi\colon N \to K_2$ is a monomorphism and that (N, ϕ) is an upper bound for $\mathscr{C}$. Thus, by Zorn's lemma, S has a maximal element (M, θ). We must show that $M = L$.

If not, there exists $\alpha \in L\backslash M$. α is algebraic over M: let m be its minimal polynomial over M. Then $\theta(m)$ splits over K_2, since K_2 is algebraically closed. Let

$$\theta(m) = (x - \beta_1)\dots(x - \beta_r).$$

Then $\theta(m)(\beta_1) = 0$, and so by Theorem 6.4 there exists a monomorphism $\theta_1\colon M(\alpha) \to K_2$ with $\theta_1|_M = \theta$. This contradicts the maximality of (M, θ). If L is finite or countable, the result follows by a standard induction argument, without appealing to the axiom of choice.

We are now in a position to show that an algebraic closure is essentially unique.   □

Theorem 18.4 *Suppose that* (i_1, K, L_1) *and* (i_2, K, L_2) *are two algebraic closures for* K. *Then there exists an isomorphism* $j\colon L_1 \to L_2$ *such that* $i_2 = ji_1$.

Proof   By Theorem 18.3 there exists a monomorphism $j\colon L_1 \to L_2$ such that $i_2 = ji_1$.

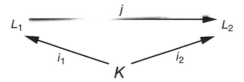

We now use Theorem 18.1. If f is irreducible over $K[x]$, $i_1(f)$ splits over L_1, and so $i_2(f)$ splits over $j(L_1)$. As $(i_2, K, j(L_1))$ is algebraic, $(i_2, K, j(L_1))$ is an algebraic closure for K. Now $L_2 : j(L_1)$ is algebraic, as (i_2, K, L_2) is, and so $L_2 = j(L_1)$, by Theorem 18.1(iii).

In future, if K is any field, we shall denote by $\overline{K} : K$ any algebraic closure of K.   □

Exercises

18.4   What is the algebraic closure of $\mathbb{Q}$ (as a subfield of $\mathbb{C}$)?

18.5   Show that an algebraically closed field must be infinite.

18.6   Suppose that $K(\alpha) : K$ is a simple extension and that α is transcendental over K. Show that $K(\alpha)$ is not algebraically closed.

18.7   Suppose that K is a countable field. Show how to construct an algebraic closure, by successively constructing splitting fields of the (countably many) polynomials in $K[x]$. Is your construction less fallacious than the 'fallacious proof' of Theorem 18.2?

18.8   Suppose that $L : K$ is algebraic. In what sense is it true that $\overline{L} = \overline{K}$?

18.4 Conclusions

We have now achieved what we set out to do. Some comments are in order. First, the proof of Theorem 18.2 is quite difficult, More to the point, it is quite different from the very special construction of the complex field $\mathbb{C}$. Here, the hard work is constructing the real number field $\mathbb{R}$ from the rational field $\mathbb{Q}$. $\mathbb{C} : \mathbb{R}$ is then a splitting field extension for the polynomial $x^2 + 1$, which is irreducible over $\mathbb{R}$. It is then remarkably the case that all polynomials over $\mathbb{R}$ split over $\mathbb{C}$. The complex field is a very special one!

Second, the proof uses the axiom of choice in an essential way. This suggests that the theorem should only be used when it is necessary to do so.

Third, the existence of an algebraic closure, and the extension theorem (Theorem 18.3) provide a useful framework in which to work. If one uses this, the theory can be developed more simply in a few places. But the use of the axiom of choice seems too big a price to pay: for this reason we have not used algebraic closures in the development of the theory.

19

Transcendental Elements and Algebraic Independence

19.1 Transcendental Elements and Algebraic Independence

In this chapter we leave the study of algebraic extensions, and consider problems concerning transcendence.

Suppose that $L : K$ is an extension and that $\alpha \in L$. Recall that α is transcendental over K if the evaluation map $E_\alpha : K[x] \to L$ is one-to-one; that is, α satisfies no non-zero polynomial relation with coefficients in K.

Theorem 19.1 *Suppose that $L : K$ is an extension and that $\alpha \in L$ is transcendental over K. Then the evaluation map E_α can be extended uniquely to an isomorphism F_α from the field $K(x)$ of rational expressions in x over K onto the field $K(\alpha)$.*

Proof The proof should be quite obvious: here are the details.

Remember that the field $K(x)$ is obtained by considering an equivalence relation on $K[x] \times (K[x])^*$ (see Section 2.4).

Suppose that $(f, g) \in K[x] \times (K[x])^*$. As α is transcendental over K, $g(\alpha) \neq 0$, and we can define $G_\alpha(f, g) = f(\alpha)(g(\alpha))^{-1}$. If $(f, g) \sim (f', g')$ then $fg' = f'g$ in $K[x]$, so that $f(\alpha)g'(\alpha) = f'(\alpha)g(\alpha)$ and $G_\alpha(f, g) = G_\alpha(f', g')$. Thus G_α is constant on equivalence classes: we can therefore define $F_\alpha(f/g) = G_\alpha(f, g)$. It is straightforward to verify that F_α is a ring homomorphism. Since $F_\alpha(x) = E_\alpha(x) = \alpha$, $F_\alpha(K(x)) \supseteq K(\alpha)$. In contrast, if $f/g \in K(x)$, $F_\alpha(f/g) = f(\alpha)(g(\alpha))^{-1} \in K(\alpha)$, and so $F_\alpha(K(x)) = K(\alpha)$. Finally if F'_α is another monomorphism which extends E_α, the set

$$\{r \in K(x): F_\alpha(r) = F'_\alpha(r)\}$$

is a subfield of $K(x)$ which contains $K[x]$; it must therefore be the whole of $K(x)$, and so F_α is unique.

We now generalize the idea of a transcendental element. Suppose that $L : K$ is an extension and that $A = \{\alpha_1, \ldots, \alpha_n\}$ is a finite subset of L (where $\alpha_1, \ldots, \alpha_n$ are distinct). Any element f of $K[x_1, \ldots, x_n]$ can be written in the form

$$f = \sum_{j=1}^{m} k_j x_1^{d_{1,j}} \ldots x_n^{d_{n,j}}$$

where $k_j \in K$ for $1 \leqslant j \leqslant m$, and $d_{i,j}$ is a non-negative integer for $1 \leqslant i \leqslant n$, $1 \leqslant j \leqslant m$. We define the *evaluation map* E_A from $K[x_1, \ldots, x_n]$ into L by setting

$$E_A(f) = \sum_{j=1}^{m} k_j \alpha_1^{d_{1,j}} \ldots \alpha_n^{d_{n,j}}.$$

It is easy to see that E_A is a ring homomorphism. We shall frequently write $E_A(f)$ as $f(\alpha_1, \ldots, \alpha_n)$.

We say that A is *algebraically independent* over K if E_A is one-to-one: that is, there is no polynomial relation, with coefficients in K, between the elements $\alpha_1, \ldots, \alpha_n$. Thus a one-point set $\{\alpha\}$ is algebraically independent over K if and only if α is transcendental over K.

We say that an arbitrary subset S of L is algebraically independent over K if each of its finite subsets is algebraically independent over K.

The proof of the next result is exactly similar to the proof of Theorem 19.1: this time we omit the details.   □

Theorem 19.2 *Suppose that $L : K$ is an extension and that $A = \{\alpha_1, \ldots, \alpha_n\}$ is algebraically independent over K. Then the evaluation map E_A can be extended uniquely to an isomorphism F_A from the field $K(x_1, \ldots, x_n)$ of rational expressions in $x_1, \ldots, x_n$ onto the field $K(\alpha_1, \ldots, \alpha_n)$.*

The next theorem is again very easy: it gives a useful practical criterion for a finite set to be algebraically independent over K.

Theorem 19.3 *Suppose that $L : K$ is an extension and that $\alpha_1, \ldots, \alpha_n$ are distinct elements of L. Let $K_0 = K$, $K_i = K(\alpha_1, \ldots, \alpha_i)$ for $1 \leqslant i \leqslant n$. Then $A = \{\alpha_1, \ldots, \alpha_n\}$ is algebraically independent over K if and only if α_i is transcendental over K_{i-1}, for $1 \leqslant i \leqslant n$.*

Proof Suppose that α_i is algebraic over K_{i-1}. Thus

$$f(\alpha_i) = k_0 + k_1 \alpha_i + \cdots + k_r \alpha_i^r = 0$$

for some non-zero f in $K_{i-1}[x]$. We can write each k_j as

$$k_j = p_j(\alpha_1, \ldots, \alpha_{i-1})(q_j(\alpha_1, \ldots, \alpha_{i-1}))^{-1}$$

where the p_j and q_j are in $K[x_1, \ldots, x_{i-1}]$ and the $q_j(\alpha_1, \ldots, \alpha_{i-1})$ are non-zero. We clear the denominators. Let

$$l_j = p_j\left(\prod_{k \neq j} q_k\right), \text{ for } 0 \leqslant j \leqslant r.$$

Then each l_j is in $K[x_1, \ldots, x_{i-1}]$ and

$$g = l_0 + l_1 x_i + \cdots + l_r x_i^r$$

is a non-zero element of $K[x_1, \ldots, x_i]$. As $g(\alpha_1, \ldots, \alpha_i) = 0$, A is not algebraically independent over K.

Conversely, suppose that $\{\alpha_1, \ldots, \alpha_n\}$ is not algebraically independent over K. There exists an index j such that $\{\alpha_1, \ldots, \alpha_{j-1}\}$ is algebraically independent over K, while $\{\alpha_1, \ldots, \alpha_j\}$ is not. Thus there exists a non-zero g in $K[x_1, \ldots, x_j]$ such that $g(\alpha_1, \ldots, \alpha_j) = 0$. Grouping terms together, we can write

$$g = k_0 + k_1 x_j + \cdots + k_r x_j^r$$

where $k_i \in K[x_1, \ldots, x_{j-1}]$ for $0 \leqslant i \leqslant r$. Let

$$h = k_0(\alpha_1, \ldots, \alpha_{j-1}) + k_1(\alpha_1, \ldots, \alpha_{j-1})x + \cdots + k_r(\alpha_1, \ldots, \alpha_{j-1})x^r.$$

Then $h \in K_{j-1}[x]$, and h is non-zero, since $\{\alpha_1, \ldots, \alpha_{j-1}\}$ is algebraically independent over K. As $h(\alpha_j) = 0$, α_j is algebraic over K_{j-1}. □

Exercises

19.1 Suppose that $L : K$ is an extension, and that $\{\alpha_1, \ldots, \alpha_s\}$ is algebraically independent over K. Show that if $\beta \in K(\alpha_1, \ldots, \alpha_s)$ and $\beta \notin K$ then β is transcendental over K.

19.2 Suppose that $K(\alpha) : K$ is a simple extension and that α is transcendental over K. Show that if τ is an automorphism of $K(\alpha)$ which fixes K then there exist a, b, c and d in K with $ad \neq bc$ such that

$$\tau(\alpha) = (a\alpha + b)/(c\alpha + d).$$

Conversely show that any such a, b, c and d determine an automorphism of $K(\alpha)$ which fixes K.

19.3 Suppose that $K(\alpha) : K$ is a simple extension and that α is transcendental over K. Let σ be the automorphism of $K(\alpha)$ which fixes K and sends α to $1/(1 - \alpha)$. Verify that σ^3 is the identity, and determine the fixed field of σ.

19.4 Suppose that $K(\alpha) : K$ is a simple extension, that α is transcendental over K, and that char K is an odd prime p. Suppose that $1 < n < p$. Let τ be the automorphism of $K(\alpha)$ which fixes K and sends α to $n\alpha$. Determine the fixed field of τ.

19.2 Transcendence Bases

We now introduce an idea which corresponds in many ways to the concept of basis of a vector space. Suppose that $L : K$ is an extension. Let $\mathscr{I}$ denote the collection of all subsets of L which are algebraically independent over K. We order $\mathscr{I}$ by inclusion. An element S of $\mathscr{I}$ which is maximal in this ordering is called a *transcendence basis* for L over K.

The next result characterizes transcendence bases.

Theorem 19.4 *Suppose that $L : K$ is an extension and that S is a subset of L. Then S is a transcendence basis for L over K if and only if S is algebraically independent over K and $L : K(S)$ is algebraic.*

Proof Suppose that S is a transcendence basis for L over K. Suppose that α is an element of L which is not in S. By the maximality of S, $S \cup \{\alpha\}$ is not algebraically independent over K, so there exist distinct $s_1, \ldots, s_n$ in S and a non-zero f in $K[x_0, \ldots, x_n]$ such that

$$f(\alpha, s_1, \ldots, s_n) = 0.$$

We can write f as

$$k_0 + k_1 x_0 + \cdots + k_j x_0^j,$$

where $k_i \in K[x_1, \ldots, x_n]$, for $0 \leqslant i \leqslant j$, and $k_j \neq 0$. Now $\{s_1, \ldots, s_n\}$ is algebraically independent over K. From this we conclude first that $j \geqslant 1$ and second that $k_j(s_1, \ldots, s_n) \neq 0$. Now

$$k_0(s_1, \ldots, s_n) + k_1(s_1, \ldots, s_n)\alpha + \cdots + k_j(s_1, \ldots, s_n)\alpha^j = 0:$$

since $k_i(s_1, \ldots, s_n) \in K(S)$, this means that α is algebraic over $K(S)$, and that $L : K(S)$ is algebraic.

Conversely, suppose that S is algebraically independent over K and that $L : K(S)$ is algebraic. If α is an element of L which is not in S, α is algebraic over $K(S)$, so there exists a non-zero

$$g = k_0 + k_1 + \cdots + k_j x^j$$

in $K(S)[x]$ such that $g(\alpha) = 0$. Each coefficient k_i involves only finitely many elements of S, and so there exists a finite subset $\{s_1, \ldots, s_n\}$ of S such that $k_i \in K(s_1, \ldots, s_n)$ for $0 \leqslant i \leqslant j$. Thus α is algebraic over $K(s_1, \ldots, s_n)$ and so $\{s_1, \ldots, s_n, \alpha\}$ is not algebraically independent over K, by Theorem 19.3. Consequently $S \cup \{\alpha\}$ is not algebraically independent over K, and S is maximal.

Just as every vector space has a basis, so does every extension $L : K$ have a transcendence basis. As in Theorem 3.9, we prove rather more. □

Theorem 19.5 *Suppose that $L : K$ is an extension, that A is a subset of L such that $L : K(A)$ is algebraic and that C is a subset of A which is algebraically independent over K. Then there exists a transcendence basis B for L over K with $C \subseteq B \subseteq A$.*

Proof   The proof is very similar in nature to the proof of Theorem 3.9. Indeed, if we replace the phrase 'linearly independent' by 'algebraically independent over K', we obtain a proof of the fact that there is a set B which is maximal among those which contain C, are contained in A and are algebraically independent over K.

The argument of Theorem 19.4 now shows that each element α of A is algebraic over $K(B)$, and so $K(A) : K(B)$ is algebraic, by Corollary 2 to Theorem 4.7. As $L : K(A)$ is algebraic, $L : K(B)$ is algebraic (Theorem 4.8) and so B is a transcendence basis for L over K, by Theorem 19.4.

Consider the extension $\mathbb{R} : \mathbb{Q}$. If S is any countable subset of $\mathbb{Q}$, $\mathbb{Q}(S)$ is countable. If $\mathbb{R} : \mathbb{Q}(S)$ were algebraic, $\mathbb{R}$ would be countable (Exercise 4.17). Thus any transcendence basis for $\mathbb{R}$ over $\mathbb{Q}$ must be uncountable.

Note that, in contrast, it follows from Theorem 19.5 that if $L : K$ is finitely generated over K then there must be a finite transcendence basis for L over K. □

Exercise

19.5   Suppose that $L : K$ is an extension, and that L is finitely generated over K. Show that the field K_a of elements of L which are algebraic over K is finitely generated over K.

19.3 Transcendence Degree

We now pursue further the parallelism with vector spaces. First we establish a version of the Steinitz exchange theorem (Theorem 3.4).

Theorem 19.6 *Suppose that $L : K$ is an extension, that $C = \{c_1, \ldots, c_r\}$ is a subset of L (with r distinct elements) which is algebraically independent over K and that $A = \{a_1, \ldots, a_s\}$ is a subset of L (with s distinct elements) such that $L : K(A)$ is algebraic. Then $r \leqslant s$, and there exists a set D, with $C \subseteq D \subseteq A \cup C$ such that $|D| = s$ and $L : K(D)$ is algebraic.*

Proof We prove this by induction on r. The result is trivially true for $r = 0$ (take $D = A$). Suppose that it is true for $r - 1$. As the set $C_0 = \{c_1, \ldots, c_{r-1}\}$ is algebraically independent over K, there exists a set D_0 with $C_0 \subseteq D_0 \subseteq A \cup C_0$ such that $|D_0| = s$ and $L : K(D_0)$ is algebraic. By relabelling A if necessary, we can suppose that

$$D_0 = \{c_1, \ldots, c_{r-1}, a_r, a_{r+1}, \ldots, a_s\}.$$

As $L : K(D_0)$ is algebraic, c_r is algebraic over $K(D_0)$. As $\{c_1, \ldots, c_r\}$ is algebraically independent over K, c_r is transcendental over $K(c_1, \ldots, c_{r-1})$ (by Theorem 19.3). Thus $s \geqslant r$. Also, by Theorem 19.3 again,

$$E = \{c_1, \ldots, c_{r-1}, c_r, a_r, a_{r+1}, \ldots, a_s\}$$

is algebraically dependent over K. Using Theorem 19.3 once more, and using the fact that $\{c_1, \ldots, c_r\}$ is algebraically independent over K, we conclude that there exists t, with $r \leqslant t \leqslant s$, such that a_t is algebraic over $K(c_1, \ldots, c_r, a_r, \ldots, a_{t-1})$. Let $D = \{c_1, \ldots, c_r, a_r, \ldots, a_{t-1}, a_{t+1}, \ldots, a_s\}$. Then a_t is algebraic over $K(D)$, and so $K(E) : K(D)$ is algebraic. As $E \supseteq D_0$, $L : K(E)$ is algebraic, and so $L : K(D)$ is algebraic, by Theorem 4.8. As $C \subseteq D \subseteq A \cup C$ and $|D| = s$, this completes the proof.   □

Corollary *If $L : K$ is an extension, and S and T are two transcendence bases for L over K, then either S and T are both infinite or S and T have the same finite number of elements.*

If an extension $L : K$ has a finite transcendence basis, we define its *transcendence degree* to be the number of elements in the transcendence basis; otherwise we define the transcendence degree to be ∞.

19.4 The Tower Law for Transcendence Degree

Suppose that $M : L$ and $L : K$ are extensions. How is the transcendence degree of $M : K$ related to the transcendence degrees of $M : L$ and $L : K$?

Theorem 19.7 *Suppose that $M : L$ and $L : K$ are extensions, that A is a subset of L which is algebraically independent over K and that B is a subset* *of M which is algebraically independent over L. Then $A \cup B$ is algebraically independent over K.*

Proof Let C be a finite subset of $A \cup B$. We can write

$$C = \{\alpha_1, \ldots, \alpha_r, \beta_1, \ldots, \beta_s\}$$

with $\alpha_i \in A$, $\beta_j \in B$. By Theorem 19.3, α_i is transcendental over $K(\alpha_1, \ldots, \alpha_{i-1})$ for $1 \leqslant i \leqslant r$ and β_j is transcendental over $L(\beta_1, \ldots, \beta_{j-1})$ for $1 \leqslant j \leqslant s$, and so β_j is transcendental over $K(\alpha_1, \ldots, \alpha_r, \beta_1, \ldots, \beta_{j-1})$ for $1 \leqslant j \leqslant s$. Thus C is algebraically independent over K, by Theorem 19.3. Since this holds for any finite subset of $A \cup B$, $A \cup B$ is algebraically independent over K. $\square$

Theorem 19.8 *Suppose that $M : L$ and $L : K$ are extensions, that A is a transcendence basis for L over K and that B is a transcendence basis for M over L. Then $A \cup B$ is a transcendence basis for M over K.*

Proof By Theorems 19.4 and 19.7 it is enough to show that $M : K(A \cup B)$ *is algebraic.*

Since A is a transcendence basis for L over K, $L : K(A)$ is algebraic. Since $K(A) \subseteq K(A \cup B)$, it follows that $K(A \cup B)(L) : K(A \cup B)$ is algebraic. As $K(A \cup B)(L) = L(B)$, this means that $L(B) : K(A \cup B)$ is algebraic. But B is a transcendence basis for M over L, and so $M : L(B)$ is algebraic. Thus $M : K(A \cup B)$ is algebraic, by Theorem 4.8. $\square$

Corollary *If $M : L$ and $L : K$ are extensions, the transcendence degree of $M : K$ is the sum of the transcendence degrees of $M : L$ and $L : K$.*

19.5 Lüroth's Theorem

Suppose that $L : K$ is a finitely generated extension which has transcendence degree r. If $\alpha_1, \ldots, \alpha_r$, is a transcendence basis for L over K, then $L : K(\alpha_1, \ldots, \alpha_r)$ is finite. If we can find a transcendence basis $\alpha_1, \ldots, \alpha_r$ for L over K such that $L = K(\alpha_1, \ldots, \alpha_r)$, then we say that L is *purely transcendental* over K. Even in particular cases, it is not easy to determine whether a finitely generated extension is purely transcendental or not (see Exercises 19.5 and 19.6 below). There is, however, one case where the problem can be solved in a straightforward way. The proof involves polynomials in two variables: first make sure that you are familiar with the contents of Section 2.10.

Theorem 19.9 (Lüroth's theorem) *Suppose that $K(t) : K$ is a simple extension and that t is transcendental over K. If L is a subfield of $K(t)$ containing K then $L : K$ is a simple extension.*

Proof We clearly need only consider the case where L is different from both K and $K(t)$. If $s \in L\backslash K$, we can write $s = p(t)/q(t)$ where p and q are non-zero polynomials in $K[x]$. Then $q(t)s - p(t) = 0$, and t is algebraic over L. Let m be the minimal polynomial of t over L. We can consider m as an element of $K(t)[x]$; by Theorem 2.31, there exists β in $K(t)$ such that $\beta m = f$, where

$$f = a_0(t) + a_1(t)x + \cdots + a_n(t)x^n$$

is a primitive polynomial in $K[t][x]$. Note that

$$n = \text{degree } m = [K(t) : L].$$

Since m is monic, $\beta = a_n(t)$ and the terms $a_i(t)/a_n(t)$ are all in L; in contrast, they are not all in K, since t is transcendental over K. There therefore exists i, with $0 \leqslant i < n$, such that $u = a_i(t)/a_n(t) \in L\backslash K$. We can write u as $g(t)/h(t)$ where g and h are relatively prime polynomials in $K[t]$.

$$\text{Let } r = \max(\text{degree } g, \text{degree } h).$$

Then $[K(t) : K(u)] = r$. As $K(u) \subseteq L$, this means that $r \geqslant n$. It also means that it is sufficient to show that $r \leqslant n$, for then it follows that $L = K(u)$.

We now consider the expression

$$l = g(t)h(x) - h(t)g(x).$$

As g and h are relatively prime, l is non-zero. Now $(h(t))^{-1}l \in L[x]$, and $(h(t))^{-1}l$ has t as a root: thus m divides $(h(t))^{-1}l$ in $L[x]$. This implies that f divides l in $K(t)[x]$. As f is primitive in $K[t][x]$, it follows from Corollary 1 to Theorem 2.33 that f divides l in $K[t][x]$. Thus there exists j in $K[t][x]$ such that $l = fj$.

We can consider f, l and j either as elements of $K[t][x]$ or as elements of $K[x][t]$: let us denote the degree in x by $\deg_x$ and the degree in t by $\deg_t$.

Now $\deg_t(l) \leqslant r$ and $\deg_t(f) \geqslant r$: since $l = fj$, $\deg_t(l) = \deg_t(f) = r$ and $\deg_t(j) = 0$. In other words, we can consider j as an element of $K[x]$. In particular, this means that j is primitive in $K[t][x]$, and so by Theorem 2.32 $l = fj$ is primitive in $K[t][x]$. As l is skew-symmetric in t and x, this implies that l is primitive in $K[x][t]$. But $j \in K[x]$, and j divides l; thus j must be a unit in $K[x]$, and so $j \in K$. Consequently

$$n = \deg_x(f) = \deg_x(l) = \deg_t(l) = \deg_t(f) \geqslant r,$$

and the theorem is proved. □

Does Lüroth's theorem extend to purely transcendental extensions of higher transcendence degree? It can be shown that if t_1 and t_2 are algebraically independent over an algebraically closed field K, and M is a subfield of $K(t_1, t_2)$ for which $K(t_1, t_2) : M$ is finite and separable, then M is purely transcendental over K. It can also be shown that a corresponding result does not hold for extensions of transcendence degree 3. These results involve polynomials in several variables in a more fundamental way than does Lüroth's theorem. The results really belong to algebraic geometry: they are discussed, for example, in the book by Hartshorne.[1]

Exercises

19.6 Suppose that $K(x, y) : K$ is an extension with x transcendental over K and $x^2 + y^2 = 1$. Show that $K(x, y) = K(u)$, where $u = (1 + y)/x$.

19.7 Suppose that $n \geqslant 3$, that $K(x, y) : K$ is an extension with x transcendental over K and $x^n + y^n = 1$ and that char K does not divide n. Suppose if possible that $K(x, y) = K(s)$.

   (i) Show that there are relatively prime polynomials f, g and h in $K[x]$ such that max (degree f, degree g, degree h) $\geqslant 1$ and $f^n + g^n = h^n$.
   (ii) Show that

$$f^{n-1} \mid (hDg - gDh) \quad \text{and} \quad g^{n-1} \mid (hDf - fDh),$$

   and show (by considering degrees) that this is not possible.

[1] R. Hartshorne, *Algebraic Geometry*, Springer-Verlag, 1977.

20

Generic and Symmetric Polynomials

20.1 Generic and Symmetric Polynomials

When we considered cubic polynomials, we saw that a variety of possibilities can arise. Some depend on the original field K: whether or not K contains cube roots of unity, for example. Others depend on special relationships between the coefficients: these can confuse the issue, and it is sensible to consider polynomials where this cannot happen.

Suppose that K is a field. Let $K(a_1, \ldots, a_n) : K$ be an extension such that $\{a_1, \ldots, a_n\}$ is algebraically independent over K. Then the generic (monic) polynomial of degree n over K is the polynomial

$$x^n - a_1 x^{n-1} + \cdots + (-1)^n a_n.$$

Note that this is an element of $K(a_1, \ldots, a_n)[x]$, and *not* an element of $K[x]$. Note also that, since $\{a_1, \ldots, a_n\}$ is algebraically independent over K, there is no relationship between the coefficients: they are quite general.

We can also consider polynomials with general roots. Let $K(t_1, \ldots, t_n) : K$ be another extension such that $\{t_1, \ldots, t_n\}$ is algebraically independent over K. Then we consider the polynomial

$$f = (x - t_1) \ldots (x - t_n).$$

Again, this is an element of $K(t_1, \ldots, t_n)[x]$, and *not* an element of $K[x]$. We can write

$$f = (x - t_1) \ldots (x - t_n) = x^n - s_1 x^{n-1} + \cdots + (-1)^n s_n,$$

186

where

$$s_1 = t_1 + \cdots + t_n,$$

$$s_2 = \sum_{1 \leqslant i < j \leqslant n} t_i t_j,$$

$$\vdots$$

$$s_n = t_1 t_2 \ldots t_n.$$

The expressions $s_1, \ldots, s_n$, considered as elements of $K[t_1, \ldots, t_n]$, are the *elementary symmetric polynomials in n variables*.

Suppose now that σ is a permutation of $\{1, \ldots, n\}$. Then σ determines an automorphism of $K(t_1, \ldots, t_n)$:

$$\text{if } \alpha = \frac{h(t_1, \ldots, t_n)}{g(t_1, \ldots, t_n)} \text{ then } \sigma(\alpha) = \frac{h(t_{\sigma(1)}, \ldots, t_{\sigma(n)})}{g(t_{\sigma(1)}, \ldots, t_{\sigma(n)})}.$$

Let G be the group of all such automorphisms, and let L be the fixed field of G. Then, by Theorem 9.3, $K(t_1, \ldots, t_n) : L$ is a Galois extension, with Galois group G.

Theorem 20.1 $L = K(s_1, \ldots, s_n)$.

Proof We can consider f as an element of $K(s_1, \ldots, s_n)[x]$. Then $K(t_1, \ldots, t_n) : K(s_1, \ldots, s_n)$ is a splitting field extension for f, and so $[K(t_1, \ldots, t_n) : K(s_1, \ldots, s_n)] \leqslant n!$, by Theorem 6.3. But clearly $K(s_1, \ldots, s_n) \subseteq L$ and $[K(t_1, \ldots, t_n) : L] = n!$: it therefore follows that $L = K(s_1, \ldots, s_n)$. $\square$

Corollary f is irreducible over $K(s_1, \ldots, s_n)$ and $\Gamma(f) \cong \Sigma_n$, the group of permutations of $\{1, \ldots, n\}$.

Theorem 20.2 *The elementary symmetric polynomials $s_1, \ldots, s_n$ are algebraically independent over K.*

Proof The transcendence degree of $K(t_1, \ldots, t_n) : K$ is n. As $K(t_1, \ldots, t_n) : K(s_1, \ldots, s_n)$ is algebraic, $\{s_1, \ldots, s_n\}$ contains a transcendence basis for $K(t_1, \ldots, t_n)$ over K, by Theorem 19.5. This must have n elements, by the corollary to Theorem 19.6, and so it must be the whole of $\{s_1, \ldots, s_n\}$. $\square$

By Theorem 19.2, this means that there is an isomorphism of $K(a_1, \ldots, a_n)$ onto $K(s_1, \ldots, s_n)$, which sends a_i to s_i (for $1 \leqslant i \leqslant n$) and which sends the generic polynomial $x^n - a_1 x^n + \cdots + (-1)^n a_n$ to f. Thus f has the same properties as the generic polynomial. Summing up:

Theorem 20.3 *The generic polynomial*

$$x^n - a_1 x^{n-1} + \cdots + (-1)^n a_n$$

is irreducible over $K(a_1, \ldots, a_n)$. *It is separable, and its Galois group is isomorphic to* Σ_n. *It is solvable by radicals if and only if* $n \leqslant 4$.

We say that a polynomial f in $K[t_1, \ldots, t_n]$ is *symmetric* if $\sigma(f) = f$ for each σ in G. It follows from Theorem 20.1 that if f is a symmetric polynomial then f can be expressed as a *rational* expression in $s_1, \ldots, s_n$. This is clearly not a satisfactory result: let us improve on it.

Theorem 20.4 *If f is a symmetric polynomial in $K[t_1, \ldots, t_n]$, there exists a unique g in $K[x_1, \ldots, x_n]$ such that*

$$f(t_1, \ldots, t_n) = g(s_1, \ldots, s_n).$$

This can also be proved directly:

Proof We consider $K[t_1, \ldots, t_n]$ as a vector space over K. A *monomial* is an element of the form $t_1^{k_1} \ldots t_n^{k_n}$; the monomials form a basis for $K[t_1, \ldots, t_n]$. We give the monomials their lexicographic order: if $u = t_1^{k_1} \ldots t_n^{k_n}$ and $v = t_1(l_1) \ldots t_n^{l_n}$ are two distinct monomials, let j be the least integer for which $k_j \neq l_j$. We then set $v < u$ if $l_j < k_j$: this is a total order on the set M of monomials. If $f \in K[t_1, \ldots, t_n]$ and $f \neq 0$, we can write $f = \sum_{r=1}^{s} a_r m_r$ where $a_r \in K \setminus \{0\}$ for each r and $m_1, \ldots, m_s$ are distinct monomials, arranged in decreasing order. Then $a_1 m_1$ is the *leading term* of f.

If f is a symmetric polynomial with leading term $a_1 t_1^{k_1} \ldots t_n^{k_n}$ then clearly $k_1 \geq k_2 \geq \cdots \geq k_n$. Let $\tilde{f} = a_1 s_1^{k_1 - k_2} s_2^{k_2 - k_3} \ldots s_{n-1}^{k_{n-1} - k_n} s_n^{k_n}$. Then f and $\tilde{f}$ have the same leading term; if $f - \tilde{f}$ has leading term $b_1 t_1^{l_1} \ldots t_n^{l_n}$ then $l_1 < k_1$, so that a straightforward induction shows that there is a polynomial g such that $f(t_1, \ldots, t_n) = g(s_1, \ldots, s_n)$, and that g is unique.   □

Corollary *Suppose that*

$$h = x^n - a_1 x^{n-1} + \cdots + (-1)^n a_n$$

is a monic polynomial in $K[x]$, with roots $\alpha_1, \ldots, \alpha_n$ in some splitting field extension. If f is a symmetric polynomial in $t_1, \ldots, t_n$ then

$$f(\alpha_1, \ldots, \alpha_n) = g(a_1, \ldots, a_n).$$

Appendix: The Axiom of Choice

Algebra is very largely concerned with considering finitely many operations on finitely many objects. When countably many objects are concerned, we can usually use mathematical induction. There are however one or two occasions when we consider sets which may not be countable; here we appeal to the axiom of choice.

A.1 The Axiom of Choice

In its simplest form, the axiom of choice can be expressed as follows. Suppose that $\{E_\alpha\}_{\alpha \in A}$ is an indexed family of sets, and that each of the sets E_α is not empty. Then the axiom of choice says that the Cartesian product $\prod_{\alpha \in A}(E_\alpha)$ is also non-empty: that is, there exists an element $(c_\alpha)_{\alpha \in A}$ in the product. In these terms, the axiom of choice may seem rather self-evident: each E_α is not empty, and so we can find a suitable c_α. The point is that we want to be able to make this choice simultaneously. The more one thinks about it, the more one discovers that this is a rather strong requirement. The axiom of choice is a genuine axiom of set theory; most mathematicians accept it and use it, as we certainly shall, but there are those who do not. Arguments which use the axiom of choice, or one of its equivalents, have a character of their own. You should avoid using it unless it is really necessary.

A.2 Zorn's lemma

In the form in which we have described it, the axiom of choice is a rather unwieldly tool. There are many statements which are equivalent to the axiom of choice (in the sense that they can be deduced using the axiom of choice, and the axiom of choice can be deduced from them). Thus an equivalent statement is that every set can be 'well ordered' (see Exercise A.2 for a definition): this is fundamental to the theory of ordinals, and leads to the idea of 'transfinite induction'. Some 90 years ago, this was the most popular and effective way of using the axiom of choice.

More recently, it has become customary to use another equivalent of the axiom of choice, namely Zorn's lemma. This is a technical result concerning partially ordered sets, which proves to be rather simple to use in practice. In order to state it, we need to say something about partially ordered sets.

A relation $\leqslant$ on a set S is said to be a *partial order* if

(a) $x \leqslant x$ for all x in S,
(b) if $x \leqslant y$ and $y \leqslant z$ then $x \leqslant z$ and
(c) if $x \leqslant y$ and $y \leqslant x$ then $x = y$.

For example, if S is a collection of subsets of a set X, the relation $E \leqslant F$ if $E \subseteq F$ is a partial order on S ('ordering by inclusion').

A partially ordered set S is *totally ordered* if any two elements can be compared: if x and y are in S then either $x \leqslant y$ or $y \leqslant x$. A non-empty subset C of a partially ordered set S is a *chain* if it is totally ordered in the ordering inherited from S.

If A is a subset of a partially ordered set S, an element x of S is an *upper bound* for A if $a \leqslant x$ for each a in A. An upper bound may or may not belong to A: A may have many upper bounds, or none at all. For example, let S be the collection of finite subsets of an infinite set X, ordered by inclusion. S itself has no upper bound, and a subset of S has an upper bound in S if and only if it is finite.

Finally, an element x of a partially ordered set S is *maximal* in S if, whenever $x \leqslant y$, we must have $x = y$. In other words x is maximal if there are no larger elements. A maximal element need not be an upper bound for S; there may be other elements which cannot be compared with x. S may well have many maximal elements.

We are now in a position to state Zorn's lemma.

Zorn's lemma *Suppose that S is a partially ordered set with the property that every chain in S has an upper bound. Then S has at least one maximal element.*

We shall not show how to deduce this from the axiom of choice. A proof can be found in Garling.[1] After working through the proof of Theorem A.1, you should be able to tackle Exercises A.1 and A.2.

A.3 The Countable Case

There is one case where Zorn's lemma does not require the axiom of choice.

Theorem A.1 *Suppose that X is a countable set, and that L is a non-empty set of non-empty subsets of X, ordered by inclusion. If every increasing sequence in L has an upper bound, then L has at least one maximal element.*

Proof   The proof is trivially true if X is finite. Suppose that X is countably infinite; let $\{x_n : n \in \mathbb{N}\}$ be an enumeration of X. If a is a non-empty subset of X, let $g(a) = \inf\{n : x_n \in a\}$; if B is a non-empty set of non-empty subsets of X, let $h(B) = \inf\{g(a) : a \in B\}$. There exists $l_0 \in L$ such that $g(l_0) = h(L)$. If l_0 is not maximal, let $L_1 = \{l \setminus l_0 : l \in L, l > l_0\}$; there exists $l_1 \in L$ such that $g(l_1 \setminus l_0) = h(L_1)$. If l_1 is maximal, we are finished. If not, we continue to repeat the procedure. Then either we reach a maximal element after a finite number of steps, or the procedure continues indefinitely.

[1] D.J.H. Garling, *A Course in Mathematical Analysis*, Vol. 1, Cambridge University Press, 2013.

If the procedure continues indefinitely, we have a strictly increasing sequence $(l_i)_{i=0}^{\infty}$ in L, such that $(g(l_i \setminus l_{i-1}))_{i=1}^{\infty}$ is a strictly increasing sequence in $\mathbb{N}$. Thus $(l_i)_{i=0}^{\infty}$ has an upper bound l_{∞}. We show that l_{∞} is maximal. For if not, and $l' > l_{\infty}$ then $g(l' \setminus l_{\infty}) < g(l_n \setminus l_{n-1})$ for some n, giving a contradiction.   □

Exercises

A.1   Show that the axiom of choice is a consequence of Zorn's lemma. (*Hint:* Suppose that $\{E_\alpha\}_{\alpha \in A}$ is a family of non-empty sets. Take S to be all *pairs* $(B, (c_\beta)_{\beta \in B})$ where B is a subset of A, and $c_\beta \in E_\beta$ for $\beta \in B$. Partially order S by setting $(B, (c_\beta)_{\beta \in B}) \leqslant (C, (c'_\gamma)_{\gamma \in C})$ if $B \subseteq C$ and $c_\beta = c'_\beta$ for $\beta \in B$. Show that every chain has an upper bound, and that every maximal element has A as its first term.)

A.2   A total order on a set S is said to be a 'well-ordering' if every non-empty subset of S has a least element. Use Zorn's lemma to show that every non-empty set can be given a well-ordering. (*Hint:* Define a partial order $\prec$ on all pairs $(T, \leqslant)$, where T is a subset of S, and $\leqslant$ is a well-ordering on T, by saying that

$$(T_1, \leqslant_1) \prec (T_2, \leqslant_2)$$

if first $T_1 \subseteq T_2$, second the orderings $\leqslant_1$ and $\leqslant_2$ coincide on T_1 and third

$$P_i = \{x \in T_2 : x \leqslant_2 t\}$$

is contained in T_1 whenever t is in T_1. Apply Zorn's lemma to this.)

A.3   Suppose that $(A, \leqslant)$ is an infinite well-ordered set with a greatest element. Show that there is a unique element a such that $\{x : x < a\}$ is infinite, while $\{x : x < b\}$ in finite for each $b < a$. Suppose that A is uncountable. Show that there is a unique element c such that $\{x : x < c\}$ is uncountable, while $\{x : x < d\}$ is countable for each $d < c$.

A.4   Suppose that $(A, \leqslant)$ and $(B, \leqslant)$ are two well-ordered sets. Show that one (and only one) of the following must occur:

(i)   there is a unique order-preserving bijection $i : A \to B$;

(ii) there exists a unique element a in A and a unique order-preserving bijection $i : \{x : x < a\} \to B$;

(iii) there exists a unique element b in B and a unique order-preserving bijection $i : A \to \{y : y < b\}$.

Index

p-group, 22

abelian, 4
ACCPI: ascending chain condition for
  principal ideals, 37
algebra, fundamental theorem of, 85
algebraic element, 74
algebraic extension, 77
algebraic geometry, 184
algebraic number, 78
algebraically closed, 170
algorithm
  for factorizing polynomials, 49
associate, 37
automorphism, 5
  inner, 5
automorphism of a field, 80
axiom of choice
  equivalence to the existence of maximal
    proper ideals, 35

basis
  existence of, 190
  for a vector space, 56
  transcendence, 180
bijection, 4

Cantor, 74
centralizer, 5
centre, 5
centre of a group, 140
chain, 190
character, 61
characteristic *n*, 29
characteristic 0, 29
characteristic polynomial, 65

choice, axiom of
  equivalence to the existence of maximal
    proper ideals, 35
closure
  normal, 100
column, 62
commutative, 4
commutative ring with a, 126
commutative ring with a 1, 25
commutator, 4
commute, 4
conjugacy class, 6
conjugate, 5
conjugation, 5
constant, 26
constructible, 147
content of a polynomial, 41
coset
  right, left, 6
cubic polynomial, 47
cubic resolvent, 154
cycle
  of type *j*, 12
cycle type, 12
cyclic extension, 135

Dedekind's lemma, 62
degree, 26
  of an extension, 71
dependent, linear, 56
derivative, 106
determinant, 62, 76, 116, 168
differentiation, 106
dihedral group, 13
dimension, of a vector space, 55, 58
discriminant, 125, 144

eigenvalue, 65
eigenvector, 65
Eisenstein's criterion for irreducibility, 94, 118
elementary symmetric polynomial, 187
epimorphism, 5
Euclid's algorithm, 42
Euclidean domain, 41
Euclidean function, 41
evaluation map, 30, 74
exchange theorem, Steinitz, 57, 181
extension
    algebraic, 77
    degree of, 71
    field, 70
    finite, 71
    generated by a set, 73
    infinite, 71
    Kummer, 137
    normal, 98
    of monomorphisms, 89
    separable, 103
    simple, 72
    splitting field, 86, 98

factorization, 36
Fermat number, 146
Fermat prime, 146
field, 32
    extension, 70
    finite, 60
    fixed, 112
    infinite, 60
    of fractions, 69
    prime subfield of, 48
    splitting, 86
field of fractions, 33
finite extension, 71
finite field, 60
finite-dimensional vector space, 56
first isomorphism theorem, 8
fixed field, 112
formal power series, 27
fundamental theorem
    of algebra, 83
    of Galois theory, 118

Galois group of a polynomial, 116
Galois group of an extension, 112
Galois theory, fundamental theorem of, 118
Gauss' lemma, 45, 49

generated by a set
    extension, 73
    ideal, 27
generic polynomial, 186
geometry, algebraic, 185
group, 3
    alternating, 14
    cyclic, 5
    derived, 5
    exponent, 4
    finite, 5
    Galois, 112
    normal, 6
    order, 4
    simple, 140
    transitive group of permutations, 117

Hermite's proof that e is transcendental, 74
highest common factor, 40
Hilbert's theorem, 111
homomorphism, 5
    ring, 27

ideal, 27
    generated by a set, 27
    principal, 27
    proper, 27
image, 6, 55, 59
independent
    algebraically, 178
    linearly, 56
indeterminate, 26
index, 6
infinite extension, 71
infinite field, 60
inseparable extension, totally, 111
inseparable polynomial, 108
integers
    ring of, mod n, 28
integral domain, 29
intermediate value theorem, 118
inverse, 3, 25
irreducible, 37
irreducible polynomial
    an example, 48
    Eisenstein's criterion, 94
    localization, 51
isomorphism, 5
    ring, 27
    theorem for rings, 28

kernel, 6
  of a ring homomorphism, 27
Kronecker's factorization algorithm, 49
Kummer extension, 137

Lagrange's theorem, 7
least common multiple, 41
length, 39
Lindemann's proof that π is transcendental, 74
linear mapping, 55, 59
linear subspace, 55
linearly dependent, 56
linearly independent, 56
Liouville
  shows that transcendental numbers exist, 74
  theorem on bounded analytic functions, 85
localization principle, 52
lowest common multiple, 43
Lüroth's theorem, 183

matrix, 62, 168
  unit, 63
maximal element, 37, 190
minimal polynomial, 74
  in terms of a determinant, 76
monic, 26
monic polynomial, 74
monomorphism, 5
  extension of, 89
  of an algebraic extension, 80
  ring, 27
multiple, least common, 41
multiplicative Euclidean function, 43
multiplicity, 30
multiplicity of a root, 108, 111

natural irrationalities, theorem on, 164
Newton's identities, 124
nilpotent, 18, 26
Noetherian, 36
norm, 137
normal closure, 100
normal extension, 98
normalizer, 6
null-space, 55
number field, 33
numerical integral domain, 33

orbit, 11
order, 5
  partial, 190
  total, 190

partial order, 190
pentagon, regular, 161
permutation, 4
permutations
  transitive group of, 117
polarity, 112
polynomial
  content of, 44
  cubic, 47
  cyclotomic, 126
  elementary symmetric, 187
  equation, 69
  generic, 186
  inseparable, 108
  minimal, 74
  monic, 74
  primitive, 44
  quadratic, 47
  quartic, 153
  quintic, 161
  symmetric, 188
PP: proper principal ideals, 37
prime, 39
  subfield, 48
prime number, 37
prime ring, 29
primitive polynomial, 44
primitive roots of unity, 126
principal ideal, 28
principal ideal domain, 41
proper ideal, 27
purely transcendental, 183

quadratic equation, 69
quadratic polynomial, 47
quartic polynomial, 153
quintic polynomial, 161
quotient ring, 28

rank, 59
relatively prime, 40
Rolle's theorem, 118
root, 30
  multiplicity of, 108
  of a polynomial, 74
  of unity, 126
  primitive roots of unity, 126
  repeated, 106
row, 62
ruler and compass construction, 180

self-conjugate, 6
series
    central, 17
    derived, 17
    group, 17
    invariant, 17
    normal, 17
    subnormal, 17
    upper central, 18
signature, 14
simple, 15
simple extension, 73
simple, $\mathbb{R} : \mathbb{Q}$ is not, 73
Soicher, Leonard, 162
soluble, 18
span, 55
splitting field extension, 86, 98
stabilizer, 11, 23
Steinitz exchange theorem,
        57, 181
subfield
    prime, 48
subgroup, 4
subring, 26
Sylow p-subgroup, 22
Sylow theorems, 22
symmetric polynomial, 188
    elementary, 187

Taylor's formula, 108
total order, 190
totally inseparable, 111
tower law, 72
    for transcendence degree, 182
trace, 137

trajectory, 113
transcendence
    basis, 180
    degree, 181
transcendental element, 74, 177
transcendental, purely, 183
transitive, 12
transitive group of permutations, 117
transposition, 12
triangular, 64
triangular matrix, 63

unique factorization, 38
unique factorization domain, 39
unit, 25
unity
    primitive roots of, 126
    roots of, 126
upper bound, 190

variable, 26
vector space, 54
    basis of, 55
    finite-dimensional, 56
Vierergruppe, 15

well-ordered set, 189
Wilson's theorem, 13, 108

zero, 30
Zorn's lemma, 189
    and the construction of an algebraic closure,
        171, 174
    and the existence of proper maximal ideals,
        35